ISLANDS
Their Lives, Legends, and Lore

ISLANDS

Seon and Robert Manley

Their Lives, Legends, and Lore

Chilton Book Company   Philadelphia / New York / London

"From this land men go to another
isle that is clept Silha. In that land
is full much waste, for it is full of
serpents, of dragons and of cockodrills, that no man dare dwell there.
These cockodrills be serpents, yellow
and rayed above, and have four feet
and short thighs, and great nails as
claws or talons. . . . And there be
also many wild beasts, and namely
of elephants." (From *The Travels
of Sir John Mandeville*)

Copyright © 1970 by Seon and Robert Manley
First Edition
All rights reserved
Published in Philadelphia by Chilton Book Company
and simultaneously in Ontario, Canada,
by Thomas Nelson & Sons, Ltd.
ISBN 0-8019-5362-6
Library of Congress Catalog Card Number 76-131235
Designed by Warren Infield
Manufactured in the United States of America by
Vail-Ballou Press, Inc.

"From that country men go by the
sea ocean by an isle that is clept
Caffolos. Men of that country when
their friends be sick they hang them
upon trees, and say that it is better
that birds, that be angels of God,
eat them, than the foul worms of
the earth." (From *The Travels of
Sir John Mandeville*)

This book is for Barbara and Frank Manley and Susan and Benjamin Belcher—Islands of Friendship.

By Seon and Robert Manley

Beaches: Their Lives, Legends and Lore
The Age of the Manager: A Treasury of Our Times

By Seon Manley

My Heart's in the Heather
My Heart's in Greenwich Village
Nathaniel Hawthorne: Captain of the Imagination
Long Island Discovery
James Joyce: Two Decades of Criticism
Rudyard Kipling: Creative Adventurer
Adventures in Making: The Romance of Crafts Around the World
Teen-Age Treasury for Girls
Teen-Age Treasury of Good Humor

By Seon Manley and Gogo Lewis

To You With Love: A Treasury of Great Romantic Literature
The Oceans: A Treasury of the Sea World
Teen-Age Treasury of Our Science World
Teen-Age Treasury of the Arts
Teen-Age Treasury of Imagination and Discovery
High Adventure: A Treasury for Young Adults
Mystery! A Treasury for Young Readers
Merriment! A Treasury for Young Readers
Magic! A Treasury for Young Readers
Suspense! A Treasury for Young Adults
Polar Secrets: A Treasury of the Arctic and Antarctic
Shapes of the Supernatural
A Gallery of Ghosts

Contents

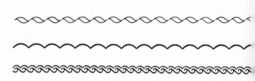

Island Fantasy

"Everyone loves an island," said the old woman sitting on her driftwood porch on Fire Island. "I came here to live when there was nothing but sand and birds. I had myself a pet gull called Bertie, and he'd eat from my hand. I gave him little treats, a clam or two from my chowder—my, he loved that!—or saltines. That bird was crazy about saltines. No, I was never lonely. I was *alone* sometimes—well, people-alone. But not world-alone. How could you be with the waves talking to you like neighbors with gossip on their tongues? And then on fine days when the water was still, it was as friendly as a blue willow tea saucer. And then of course people came regular. The old baymen would bring me a basket of snap-pers, or a real old fighting blue. Sometimes they'd sail me over to Patchogue. But I never did get my land legs. I'd hanker to be home again—besides, Bertie missed me. He liked the salt pork I brought back for chowder, though. That gull had a passion for salt pork. I cut it into bits for frying (always try your salt pork real good for my chowder), but I'd save some for Bertie and throw it in the air. Why, that bird would caw with excitement—ate dainty too, right from my fingers.

"It was my island then. Made this house with whatever the sea threw me. I always got along well with the sea—not that it couldn't be a nasty old robber at times. Stealing the sand. Now I guess it's everybody's island. Doesn't matter though; your island is always kind of inside you anyway. Besides, everybody loves an island. A pity, though, they don't understand islands. Just come and go, drop a beer can, and don't understand."

This old lady on one of our favorite islands in the entire world knew what she was talking about. The world has taken a long time learning to understand islands despite the fact that since time immemorial man has been attracted to them. It wasn't until World War II—an "island war" as it was called in the Pacific—that a serious study of islands was undertaken. At that time the Island Research Foundation asked the volatile H. L. Mencken to coin a new word for island study. He suggested "islandry" or "islandography." Both names have yet to make his, or any other, dictionary. Islandography smacks rather of the textbook—but islandry has something of the romance of island life and living, the history, the lore, the legends that an island develops. So this is an adventure into islandry—not a geography of the island world, but a portrait, not even a family portrait (there are more than half a million indexed islands)—in search of the personality of some islands, the extraordinary history of others, and the afterimage, too, of the people who loved them.

Thoreau, for example:

2

An island always pleases my imagination, even the smallest, as a small continent and integral part of the globe. I have a fancy for building my hut on one. Even a bare, grassy isle, which I can see entirely over at a glance, has some undefined and mysterious charm for me. There is commonly such a one at the junction of two rivers, whose currents bring down and deposit their respective sands in the eddy at their confluence, as it were the womb of a continent. By what a delicate and far-stretched contribution every island is made! What an enterprise of Nature thus to lay the foundations of and to build up the future continent, of golden and silver sands and the ruins of forests, with ant-like industry. . . .

The shifting islands! Who would not be willing that his house should be undermined by such a foe! The inhabitant of an island can tell what currents formed the land which he cultivates; and his earth is still being created or destroyed. There before his door, perchance, still empties the stream which brought down the material of his farm ages before, and is still bringing it down or washing it away—the graceful, gentle robber!

Thoreau's passion was for small islands—the islands of the Merrimac River and the Concord—with an occasional foray upon the Connecticut. Small river islands—Plum Island, in his day, for example, "its sand ridges scalloping along the horizon like the sea-serpent, and the distant outline broken by many a tall ship, leaning, *still*, against the sky."

3

The creation of Japan: Izanagi and Izanami stand in the clouds and create the island out of the seawater. The artist was Eitoku of the Ukiyo-E school in the late nineteenth century. (Courtesy Museum of Fine Arts, Boston, Bigelow Collection)

4

Writers, artists, and musicians have always been keen island lovers. This self-portrait of Paul Gauguin was painted in the South Seas. (Courtesy National Gallery of Art, Washington, D. C., Chester Dale Collection)

Islands have often nurtured and protected great art. This is Gal Vihare, Polonnaruwa, 133 miles from Colombo. Carved on the face of the rock are three figures of Buddha. Picture shows a standing Buddha and the enormous figure of a recumbent Buddha 44 feet in length. (Courtesy Ceylon Tourist Board)

5

Or islands with no names, or islands he named because he caught a rabbit there, or islands that stuck in his mind because of their history. Wicasuck, the island of early Indians (the Indians were great island lovers), or occasionally larger islands, Staten Island for one.

Yet I have sometimes ventured as far as to the mouth of my Snug Harbor. From an old ruined fort on Staten Island, I have loved to watch all day some vessel whose name I had read in the morning through the telegraph glass, when she first came upon the coast, and her hull heaved up and glistened in the sun, from the moment when the pilot and most adventurous news-boats met her, past the Hook, and up the narrow channel of the wide bay, till she was boarded by the health officer, and took her station at quarantine, or held on her unquestioned course to the wharves of New York. . . . And, again, in the evening of a pleasant day, it was my amusement to count the sails in sight. But as the setting sun continually brought more and more to light, still farther in the horizon, the last count always had the advantage, till, by the time the last rays streamed over the sea, I had doubled and trebled my first number; though I could no longer class them all under the several heads of ships, barks, brigs, schooners, and sloops, but most were faint generic *vessels* only. And then the temperate twilight, perchance, revealed the floating home of some sailor whose thoughts were already alienated from this American coast, and directed towards the Europe of our dreams. I have stood upon the same hilltop, when a thunder-shower, rolling down from the Catskills and Highlands, passed over the island, deluging the land; and, when it had suddenly left us in sunshine, have seen it overtake successively, with its huge shadow and dark, descending wall of rain, the vessels in the bay. Their bright sails were suddenly drooping and dark, like the sides of barns, and they seemed to shrink before the storm; while still far beyond them on the sea, through this dark veil, gleamed the sunny sails of those vessels which the storm had not yet reached. And at midnight, when all around and overhead was darkness, I have seen a field of trembling, silvery light far out on the sea, the reflection of the moonlight from the ocean, as if beyond the precincts of our night, where the moon traversed a cloudless heaven—and sometimes a dark speck in its midst, where some fortunate vessel was pursuing its happy voyage by night.

Thousands of miles away, a young poet inspired by Thoreau's *Walden*, which his father introduced him to, sang to himself: "I will arise and go now, and go to Innisfree."

6 William Butler Yeats's island, too, was a small one, a romp of an island in the native Gaelic meaning—The Island of Heather—but not only of heather, but holly, wet and shining in the mists of Lough Gill, with bracken as thick as moss, and the moss itself rich as a blanket for old Queen Maeve—and with ferns as luxurious as in any tropical rain forest.

In the old days the traveler, the adventurer was a poet too—and the wild poetry and prose filled with marvels explored incessantly the mystery of islands just beyond the horizon. There was Atlantis, of course, to which we will return—now a contemporary problem of modern archaeology and legend once again in a popular song—Donovan's *Atlantis*.

The world's first tourists, the Romans, were fascinated by the islands of the Atlantic. To them we owe the confirmed discovery of the Canary and Madeira groups, and with all the aplomb of a contemporary copywriter inducing us to make a trip to some island paradise, they called those sets of islands the Fortunate Islands after the old mythological Greek islands, the Isles of the Blest. But by the time of the Middle Ages the lore of the past had grown rusty. Many of the islands discovered by the ancients no longer appeared in the maps of the Middle Ages. All islands then became almost fanciful and abstract, even one of the most famous of all island groups, the British Isles. That great book *De imagine mundi* had only this to say about the British Isles, "Over against Spain toward the setting sun are the following islands in the ocean: Britain, England, Hibernia, Thanet—the earth of which wherever it may be carried, will destroy serpents—the thirty-three Orkneys on the Arctic Circle where the solstice occurs, Scotia, and Chile (Thule)."

The lessons of those great scholars and observers, the Greeks, had long since been forgotten by that time, although it is almost certain that by the fourth century B.C. the Greeks had explored Scandinavia, Britain, Thule and a great deal of the ocean; knew the Orkneys, the Shetlands and the Faeroes; knew, too, probably the Canaries and Ceylon.

By the time of that great Welsh writer and adventurer, Giraldus Cambrensis, the island to the west—Ireland—was gaining in fame. It was a land that flowed with milk and honey, a land of wine, a land of strange wild beasts, of poisonous yew trees, of violent winds, a land studded with lakes, an island abounding in fish. He writes, too, of the Orkneys and the Shetlands and also of Iceland and Thule. By the year 900 Greenland had been discovered, and men were hunting seals there by the twelfth century.

As the writers wrote about the north, fabulous stories entered into history—stories of great whirlpools, monsters, lands of spirits, and terrible currents. There were legends of Wineland the Good, based on Ericson's voyages to the coast of America, but simultaneously with what appeared to be true observations and memories of observations there were also fabulous islands reported: the Islands of the Gorgons; the Island of Perdita, or Lost Island, a kind of Atlantis legend. The fabulous islands, the lost islands, the islands with strange names and stranger people kept occurring in island lore.

Recently that gallant mariner, Bill Verity, sailed from Ireland to America in a boat modeled on that of St. Brendan, who supposedly made a voyage to the American coast, visiting many islands on the way. Bill Verity sailed alone, but St. Brendan himself was supposed to have taken seven companions and sailed for seven years visiting one extraordinary island after the other.

One of St. Brendan's islands was obviously a volcanic island—volcanic islands have attracted the interest of writers since the time of Pliny the Elder; Sicily and Iceland were both volcanic. In those days the volcanoes were thought of as the gates of hell, hell to the northern peoples being a cold hell. From the coast of Iceland, for example, legends developed that warned of wailing monsters all along the cliffs and crags. And some legends with a religious turn say that Judas sat alone on a rock near one of these volcanic islands on Sundays and on Christmas. The rest of the time he was imprisoned in the heart of a volcano. It was

7

Momotaro at the demon island. Japanese mural from the Kyoto school, 1852. (Courtesy Museum of Fine Arts, Boston, Bigelow Collection)

thought in those days that volcanoes were caused by the winds pent up in the bottom of the earth trying to escape.

Sophisticated early travelers—the Egyptians, Phoenicians, Greeks, Romans, and Vikings—were important island discoverers. But those great early colonists, the Polynesians, colonized most of the habitable islands of the Pacific long before Columbus. We do not know their starting point, undoubtedly someplace in Southeast Asia, but we do know that they were driven on and on in a spirit of exploration, using one island as a base and then moving to another—perhaps because of disease or an inhospitable enemy or perhaps simply because the love of the sea and the island life was deep within their consciousness. They invented the outrigger and the double canoe. According to their legends the canoes were four hundred feet long; each could carry nearly sixty men and women along with the animals and the foods that they would establish on the next new islands. The Polynesians used complex sailing charts, the sun, and the stars, and had an extraordinary knowledge of the patterns of tides around islands that only oceanographic studies in recent years have confirmed. Perhaps the Polynesians even knew those romantic circles, the Tropics of Cancer and Capricorn. In any case, by very early times they were living on the islands of Samoa, Hawaii, Tonga, and Tahiti, the Cook Islands, and probably New Zealand.

They developed extensive lore to tell how the islands were created: New Zealand, for example, was discovered first by the god Maui, the youngest of five brothers and—being the youngest—always in a little trouble with the others. Stowing away one day on a canoe belonging to his brothers, he exterted his infinite sense of command, took over the boat, and sailed into unknown waters. He found a perfect place to fish, but he had no fishhook and his brothers would not share theirs. So, being a practical boy, he used the jawbone of his grandmother, which he just happened to have with him, tossed it into the sea and came up with the North Island of New Zealand. Today his name lingers on the island of Maui in the Hawaiian group.

Just as Thoreau was delighted to give a name to some small river island that he came across, all of the great island travelers have bestowed names upon the islands of the world—none more richly named by a variety of peoples, especially the Polynesians, than the islands in the Pacific, each nationality using a nomenclature peculiar to it.

There are infinite numbers of tales of the disappearing and reappearing islands of the world. Sometimes the legends were based on sensible observations: islands that floated away, for example, and disappeared in the tide, carried away by the currents, or vegetation that disappeared in an appropriate season, only to appear once again from the seed of a

9

Crystal wave cut in the profile of two overlapping billows. The upper part is engraved with a map of the Bermudas, lying amid coral reefs and vestiges of the sailing vessels lost in storms off these shores. Below, an engraved vignette suggests the charm of the islands; it includes Poseidon riding his dolphins, flora of the islands entwining lissome figures, and Warwick Fort, constructed in 1614 to control entrance to the harbor. (Courtesy Steuben Glass)

mangrove, for instance, or the remnants of a coconut. Starting with that great island explorer Magellan, navigators could sometimes pass a large collection of islands in the Pacific, noting none of them, while the next visitor might turn the prow of his boat directly to some new discovery. In the old days, of course, when telescopes were poor, when the weather was always to be reckoned with, appearing and disappearing islands or unattainable islands often turned up in the literature even of the scientists and adventurers and discoverers. A historian by the name of G. C. Henderson, author of *The Discoverers of the Fiji Islands,* had always been struck by what was true and what was not true when one was at sea, and he has some interesting observations about the old logbooks of early explorers:

Though the information given by the commanders in their charts and logbooks may be at frequent variance with actual facts, it is nevertheless a true account of what they saw. It was well known to us and was no doubt to them that the evidence of the senses is often misleading at sea; but as they plowed their way through uncharted waters, they had little or no time to make minute examinations. They had to trust very largely to their senses; the fact that their charts often differ very materially from modern charts does not mean that they were careless or incapable men. It means, rather . . . that quite frequently things are not as they seem to the voyager passing by. The sketch of the coast of Viti Levu (Fiji Islands), taken by Midshipman Langdale on the *Pioneer* in May, 1932, from Bligh's position at six a.m. on May 8th, 1789, resembles the sketch on Bligh's little chart much more closely than a modern British Admiralty Chart. The Admiralty Chart is true to fact, the sketches by Mr. Langdale and Lieutenant Bligh *are true to what they saw from that stationpoint.*

11

Crystal silhouette of a mountainous island, engraved with a view of palm-rimmed shore. Islanders paddle an outrigger canoe, weave palm leaves, carry baskets of coconuts. In the distance are a topographical map of Tahiti and a compass rose that lies behind the trees like the setting sun. (Courtesy Steuben Glass)

So if things are not what they seem, it is not surprising that from the earliest period there should occur myths and legends about appearing and disappearing islands. And some islands that turned out to be truly nonexistent were nonetheless perpetuated on charts that gave even legends credence. For example, up until the end of the nineteenth century that great mysterious island, Brazil, which was said to lie somewhere off the coast of Ireland, appeared on the charts of the British Admiralty. Some islands had a long and violent history; one such was Bermuda, the home of demons and called for years The Isle of Devils. Then, of course,

13

"Since once I sat upon a promontory,/And heard a mermaid on a dolphin's back,/ Uttering such dulcet and harmonious breath,/That the rude sea grew civil at her song." (William Shakespeare) Atlantic bottle-nosed dolphin at Gulf Islands National Seashore (proposed). (National Park Service photograph by M. W. Williams)

14 "In a coign of the cliff between lowland and highland,/At the sea-down's edge between windward and lee,/Walled round with rocks and an inland island,/The ghost of a garden fronts the sea." (A. C. Swinburne) Channel Islands, California. (Courtesy National Park Service)

Nineteenth-century Rhode Island coastal scene.

there were Antilia, the Island of the Seven Cities, and the strange, fascinating island of the Amazons, perhaps first mentioned by that indefatigable traveler, Marco Polo.

There was one island called Feminea, the home of the women, and another appropriately enough named Masculea, the home of the men. The legend went that all the women on the Isle of Women were beautiful, tall, and young, and for three months of every year, almost like sirens, they called to the sailors from passing ships. At the end of three months, plied with fruit and honey and womanhood, the men had to leave and take with them any male children who had reached early maturity. The belief in this island was so current that Columbus tried to visit it. He thought it somewhere in the area of Martinique. But the condition of his ships was such that he postponed the search—indefinitely.

Like Bermuda, an island in the St. Lawrence was referred to as The Isle of Demons. When Sieur Jean Robertval went to Canada his niece, who was traveling with him together with a nurse and a group of young soldiers, was seduced. The old nurse, the niece, and her lover were all placed ashore on The Isle of Demons, where she and she alone was eventually rescued.

15

Islands have a rich and varied folklore. Legendary ships, for example, are frequently seen from their shores. There is one off Orrs Island in Maine; there is the old legend of the Palatine of Block Island; and the same kind of story lingers on the shores of Long Island. Of all of them the legend of the Palatine ship off Block Island is one of the most fascinating.

The legend of the Palatine ship has lasted since 1732, when the boat left Rotterdam to sail for Philadelphia. She had on board three hundred and fifty wealthy Dutch Protestant Palatines who were off to the new world to seek a new way of life. They carried with them much property and many hopes. But the trip started badly; it was discoverd almost immediately that the drinking water was polluted. Soon the captain, some members of the crew, and nearly two hundred of the Palatines died. But the journey was to be further racked with trouble: It was midwinter, it took a long time to make the difficult journey, and when once the actual coast of America came into view the ship could not gain port because of stormy waters. The crew that remained hoisted black flags to identify the vessel as a pirate ship. With little water and food remaining, it was said the crew demanded absurd amounts of money for a single sip of water or a bite of a small biscuit. Eventually the crew abandoned the ship as she came into a raging blizzard off Block Island. Other legends say that there

ISLANDS

"And in that isle there is a great marvel, more to speak of than in any other part of the world. For all manner of fishes in the sea come once in the year. And they cast themselves to the sea bank of that isle so great plenty and multitude, that no man may anywhere see but fish. And there they abide three days. And every man of the country taketh of them as many as him liketh. . . . And no man knoweth the cause wherefore it may be. But they of the country say that it is for to do reverence to their king . . . as the most noble and excellent king of the world, and that is best beloved with God, as they say." (From *The Travels of Sir John Mandeville*)

16

were wreckers on the shore deliberately waiting for the poor debris that would be tossed upon the rocks. But there are equally persistent legends that the wreckers tried to do more to help than to harm. This is a common enough history on all islands throughout the world, whose inhabitants in all honesty have tried to help wrecked ships rather than become those mythological "wreckers" constantly luring ships to shore. However, one such wrecker decided to set fire to the Palatine ship because she was so dangerous to navigation. It seems rather a premature step to have taken, since there were still passengers aboard. With the lick of flames on their clothes they jumped into the December surf, many to be drowned and indeed only sixteen to survive.

One woman would not leave the ship; the water below her she thought raged more cruelly than the fire. And as the ship moved out into the sea, her screams died in the distance. The legend of the Palatine, bolstered by the fact that so many people have seen her lights, is one of the most persistent island legends of the east coast. It is still common to hear, "It is said that if you're on Block Island some evening when a storm is brewing, you will hear the shrieks of the lone woman who still sails with the Palatine." The ship was immortalized by John Greenleaf Whittier in the poem of the same name and in 1947 a memorial stone

The legendary St. Brendan conducts mass on a fish's back during his journey through innumerable islands on his way to the American coast. (From an old German engraving)

This fifteenth-century woodcut shows a town and fort in the process of construction along the seashore and the words "Insula Hyspana." This is one of the earliest woodcuts showing the New World. (Woodcut by Johann Bergmann de Olpe)

was put up in Rhode Island, but the ghost of the ship continues to sail the shore and only a few years ago lights credited to the Palatine were seen through all the southern coastal towns.

Rhode Island is rich in lore, and it has another famous ghost ship: the brig *Seabird*, famous around Newport. It was the middle of the eighteenth century when the ship first approached the shore, a beautiful sight, sails set, colors flying. A group of fishermen and farmers at the shore watched her head toward land; they were fearful that she would never get through the enormous breakers off the coast. "Go back," they shouted; "you will be dashed to pieces." But the ship moved onward, riding the breakers like a bird, and beached herself at the shore. Once on shore she was examined carefully. It was an empty ship—only a dog on the deck, a cat in the cabin, coffee on the stove. Her story is still one of mystery. How did the crew disappear? And why so rapidly? It was known that she was owned by a Newport man, that she had sailed to and was on her way back from Honduras, that she had been seen with a full crew only the day before, but within twenty-four hours she had become a dead ship, a ghost ship.

American Indians developed extensive lore about islands. Martha's Vineyard, for example, was the home of Maushope, a great giant of an Indian. He could wade across the water and reach the mainland, wetting only his feet and ankles. Only the fiercest of storms would ever make him wet his knees.

He attempted at one point to build a bridge, or a causeway, from that magnificent, magical spot of Gay Head to Cuttyhunk. He worked at it quite carefully. The rocks that you see at Gay Head are some that he abandoned; you can find some, too, at Cuttyhunk. But the bridge is, of course, unfinished. As brave and strong and tall as he was, he was attacked by giant crabs in the water, and one bit his toe, an indignity that made him give up his causeway.

They say Maushope could stand high above Gay Head, sighting whales that would come close to shore, and then with one great dip of his arm and hand pull up a leviathan from the sea, a tasty snack for the first beach bum on Martha's Vineyard. Sometimes, like the Japanese, he preferred them raw; at other times he broiled them—generally up near Devil's Den. He liked a good roaring fire and would go through the woods—woods that have long since been decimated—pulling up trees as you or I gather seaweed or pull up recalcitrant weeds in the garden. They say that some of the lignite you find around Gay Head is made out of coals from his ancient campfires. He was not much of a family man and accordingly turned his children into fish that fed the islanders for many decades.

18

"Tahitian War Galleys" by Hodges. (Courtesy National Maritime Museum)

19

Islands have always been vulnerable to invasion. This old print shows Boadicea haranguing the British tribes at the time of the Roman invasion.

ISLANDS

And how did he find this island in the first place? Did he simply wade out from the mainland until he stumbled across the beauty of Martha's Vineyard? No, they say that he found it by following a bird. In the island literature of the Pacific and of the whole world, birds often led men and gods to the islands. This particular bird left Cape Cod, and when he left he enticed children away from their pleasures on the Cape as they hid among the marsh grass or stretched their feet in the cold water spring. They used to warn the children on Cape Cod of that giant in the old days, but there were always, of course, children and young people who would not listen and who followed the bird, a curious Peter Pan, until they reached Martha's Vineyard. Why it occurred to old Maushope to follow the bird's trail, no one knew, but perhaps he just wanted a bit of a stroll and he strolled easily from the Cape until he found the Vineyard. Alas, there was nothing left of the children, but he had sense enough to enjoy his discovery.

When he arrived at the Vineyard he was a little tired, according to the legend. He sat down for a good smoke, gathering the poke of the island. He pulled it up by the handful, shoved it into his cavernous pipe, and lighted it. As he pulled on the pipe, the smoke moved like a mist covering the entire island—one of the greatest, fiercest of all fogs. And from that time on, the Indians used to say, when they could see a fog springing up across the Vineyard, "Here comes the smoke of old Maushope." That first pipe on the island, however, was particularly enjoyable. He looked around for a place to dispose of the remains of the ashes. The sea attracted his eye—it glistened in the sudden sunlight that had emerged after the fog—and there he emptied his ashes. The ashes piled up on a shallow, which grew into the island of Nantucket.

Like many Indian tribes, the tribes of Nantucket and Martha's Vineyard, so say the ancient legends, were fierce and hostile neighbors. Perhaps the mere nearness made each vulnerable to the other, but it used to be common for the tribe on Martha's Vineyard to go to Nantucket in war canoes and attack as many sleeping Indians as possible. One day a major attack was to be made well before the dawn, when their enemies would be safely asleep. They would row very quietly, barely making the sounds of mosquitoes as their oars hit the water. Those who remarked such things saw that the stars were still in the west. They were perhaps a little self-satisfied when they landed, or they would have seen, crouching on the shore, waiting warriors. Shocked and dismayed, they could not launch an organized attack and many were slain before they could pull their canoes away from shore. How did it happen? How could anyone have known about their impending attack?

A long time went by—many moons, many canoe trips—and peace eventually came to the two tribes. Then, when an old Vineyard Indian was dying, he told the story of how he had once loved a girl on Nantucket. He had loved her so much that he needed to protect her and when he heard of the attack that his tribe was planning, he got up in the middle of the night and went to Nantucket. Since he had been a small boy, he had noticed a certain neck of sand that might join the two islands if one was fast enough to run as fleetly as the wind across it. So he waited for the point of lowest tide, raced across to Nantucket, found the girl he loved and warned her of the impending attack. The race back was incredible. The waves rose higher and higher, but he did reach his own camp. He did not realize, of course, that his beloved would warn her people, and the next day he was shocked and bewildered at what he thought her treachery.

As with all great islands, there are many legends about that fine island of Mackinac. There, of course, Hiawatha was born. But more important, it was similar to the ancient Mount Olympus of the Greeks—God lived there himself. Whether he was called Gitchi Manitou, or Mitchi Manitou, it was his home, and it was on Mackinac that he first tried many of his experiments to make life easier for man and woman. There was a Garden of Eden on Mackinac and there, for the first time, he made the beaver.

21

Some say the island was really named Great Turtle—Moschenemacenut. The French, fond of turtle as they were, had difficulty with the name, and the old-word specialists say they turned it into Michillimackinac, which was, in turn, changed to Mackinac by the English. The island was created by a turtle which had been dispatched to the bottom of Lake Huron where it gulped an enormous quantity of mud and thus created the isle of Mackinac. You can find the descendants of that turtle on the island today, asleep in the sun. Leave them there. They're allowed to sleep in the sun forever as a reward from Manitou.

Other legends say that when you made the proper sacrifices to the great God, Manitou, after death you would go to dwell forever in the happy hunting ground which lay beyond the great range of the Rocky Mountains. But there were some for whom life always went wrong, who did not seem to know what were the proper sacrifices, and who would simply go on for eternity without homes, wanderers around all the Great Lakes, watched constantly by giants who never slept, and who were ten times the size of the tallest Indians. Some of these giants you can still find on Mackinac today. One, a rock called Sugar Loaf or sometimes Manitou's Wigwam, towers ninety feet high. The cave within the rock

The sirens of the Grecian islands call to Odysseus and his sailors in this vase painting.

was Manitou's sleeping place and was filled with ghosts and spirits. There are some that say the name Michillimackinac means the "place of the great dancing spirits." Beneath the island there are enormous caverns where the spirits still dance. There was one Indian who saw them dance and who had the gift of being able to put his body on and take it off at will, and that is the reason we know the story

There are other stories connected with Mackinac. One says that Manitou took a single grain of sand from the great ocean. He let it float on the waters, tending it the way a mother would tend a child. He watched it grow to the size of a young wolf that seemed propelled by perpetual energy. It ran and ran, not knowing where it ran, and finally died of exhaustion. The young wolf became the sand of the earth.

Two of the most delightful Indian legends come from Long Island:

They were beautiful lands, those lands in both Connecticut and Long Island that bordered Long Island Sound, so beautiful that the devil himself longed to possess them completely. They were owned by the Indians in those ancient days, and the Indians certainly would have no truck with the devil; though the white man might cajole them out of their land, surely they were smarter than the devil. The devil attempted all kinds of machinations, but finally he decided that he would fight for possession of the lands. The Indians of those days were great fighters, and they quickly overpowered him. The devil made his retreat to the spot which is now the head of the East River and found the tide conveniently low. He was able to step quite gracefully from island to island choosing a reef, a rock, or a spot of green land; for that reason you can still find those reefs and islands called the Devil's Stepping Stones. Once he reached Throgs Neck, a great fatigue enveloped him, and he sat down to brood. What could he do? He was a poor excuse for a devil if he could not get what he wanted. He sat down, as many of us have, on one of the Long Island beaches and saw that there were rocks all about him; on the other hand, in distant Connecticut there were no rocks at all at that time. In his rage he threw the rocks toward Connecticut in hopes of hitting both the Indians there and those on the island. His aim was poor and he made his way back to the middle of Long Island, specifically to Coram, where he pined away.

Long Island, known for years as the Island of Sweet Waters, once—says Indian mythology—suffered a drought that all but ruined the land. All the rituals and prayers could do nothing; no rain came. The salt water around the island was of no use and only in beautiful Lake Ronkonkoma was there any fresh water at all. But Lake Ronkonkoma was surrounded by spirits and one never knew whether he would be allowed to drink or not. There was, for example, a girl who rode forever in a birch canoe on

23

An old view of Michillimackinac, which the English shortened to Mackinac.

the great lake, a guardian of the water put in charge of the lake by the Great Spirit himself. She prevented all fishing there because the fish were really the souls of dead warriors who were waiting for deliverance. No, the people would not try the water of Lake Ronkonkoma, but they would pray once more to the Great Spirit and finally he heard their cry. "Take an arrow," he said, "shoot it in the air, watch where it falls, and water will be yours." No sooner had the arrow touched the ground than the sweet water of Long Island poured forth. They called that spot The Hill of God, or Manitou Hill, or later Manetta Hill.

The Indians knew the nature of their land well; it was they who supposedly made the canal from the bay to the sea at Canoe Place, and it was they who first discovered the silver that was supposed to exist at Little Neck on Long Island. The Montauks, they say, mined that land years ago. They used to carry the silver to traders for a bit of tobacco or rum. One old Indian while in his cups told of this secret hoard and let it be known to a white man named Gardiner who in thanks killed the Indian and hid him in the sand. It was supposedly Gardiner who then

The legendary Sir John Mandeville, author of a collection of remarkable early island stories.

25

Islands have often been renowned as the meeting places of history. This early print shows Magna Charta Isle and the meadow of Runnymede in England.

found the treasure only to hear a terrible wailing around him and to see a shadowy figure, the Indian he had killed. Nonetheless, he gathered up what silver he could only to discover when he took it home that it had turned to gray dust with a high luster—simply iron pyrites. Those who knew said that for the Indians it had been silver, but now there was a curse upon that land at Little Neck.

Sometimes not only the islands but lonely rocks in the water—miniature islands—will collect stories around them the way they do seaweed. On the Potomac River just above Georgetown there are, for example, the three rocks known as The Three Sisters. It was well over a cen-

Canoes of the South Pacific at the time of Captain Cook. (Courtesy British Information Service)

tury ago that three sisters took a rowboat and went out into the river, only to come to grief on the submerged reef where they all drowned. But you can still hear the sisters moan near those rocks in the Potomac, and one May night in 1889 the rocks cried so that the good people of Georgetown were awakened. They prayed for the soul that they knew would be cast upon the rocks and true to the legend the following day a young man drowned off The Three Sisters.

Why shouldn't the islands of Boston Harbor have many legends connected with them? Their shapes and names lend themselves to story: Noddle's Island, later called East Boston, shaped like an enormous polar bear with his head to the north and his feet to the east; Governor's Island shaped like a ham, Castle Island like a shoulder of pork; Apple Island, appropriately enough named for its shape; Snake Island, which in the old books used to be "likened to a kidney"; Deer Island in the shape of a whale; Thompson's Island, that of an unfledged chicken; Spectacle Island, a pair of glasses; Long Island, a high-topped military boot; Rainsford Island, a mink; Moon Island, a leg of venison; Gallop's, a leg of mutton; Lovell's, a dried salt fish; George's Island, a fortress; Pettick's Island, a young sea monster; Half Moon Island, and others resembling, as the old historian Nathaniel B. Shurtleff said, "pumpkins, grapes and nuts."

They all have their stories, some commemorated by the famous. For example, it was one George Worthylake who with his wife, Anne, and his daughter sailed one fine day to Noddle's Island where he was to be keeper of the lighthouse, only to have all three "took to Heaven by the way." They were taken to Copp's Hill and buried after they drowned, and a young man named Benjamin Franklin wrote a commemorative poem which he hawked and sold on the streets of Boston.

29

But it is Nix's Mate whose story has lingered longest. It was once not simply a small beacon for mariners but a decent-sized island. It is the geography and geology of these islands that have given them, at least in the instance of Nix's Mate, their strange stories. These islands in Boston Harbor are exceptional. They are called elliptical islands—nearly all of them, despite their varied names, having some kind of elliptical shape. Some are ellipses joined together, such as Spectacle Island; some have been cut off by the waves so that the ellipses are more difficult to recognize. All of these islands are drumlins; formed of soil and rocks by the continental ice sheet, they are like moraines. When the ice receded, it left behind loose materials which were easily eroded by the coastal waves. The erosion of Nix's Mate led to a famous legend. Back in the seventeenth century on that respectable island there were twelve good acres of land—excellent pasturage—that were given to a John Gallop, one of the best pilots of his day. Shurtleff tells us:

Mr. Gallop was a noted pilot in his day, and is said to have been better acquainted with the harbor than any other man of his time. On the fourth of September, 1633, he piloted into Boston Harbor, by a new way, probably

The prow of a Viking ship. (Courtesy University of Oldsaksamling, Oslo)

the Black Rock passage, the ship Griffin, containing, among its passengers, Rev. John Cotton, Elder Thomas Leverett, and many others, who afterwards proved to be some of the most desirable of the New England colonists. To his ability as a pilot and fisherman he added that of a good fighter; for, on one occasion, in July, 1636, he, with his two young sons, John and Samuel, and his boatman, heroically fought fourteen Indians, and rescued the body of his friend John Oldham, whom the savages had most cruelly murdered. Although Mr. Gallop lived at the north end of Boston, near the shore, where his boat could ride safely at anchor, he owned Gallop's Island, as a farm, a meadow lot on Long Island, and a pasture for his sheep upon Nix's Mate. How unkind it is, at this late time, to rob him of the good name he gave his island, and to call it, in a Frenchified manner, Galloupe's Island. . . .

There is a tradition connected with the history of this island, probably of modern date, which has no facts to sustain it. The story is, that the mate of a certain Captain Nix was executed upon it for killing his master, and that he, to the time of his death, insisted upon his innocence, and told the hangman that in proof of it the island would be washed away. As the island bore the name of Nix certainly as far back as the year 1636, and as no man was executed in the Massachusetts colony for murder or piracy so early as this, there is no good reason for believing that the name of the island originated in the manner given in the tradition. That the island in later times was used as a place for the burial of executed pirates and mutineers upon the sea is too well known to be disputed.

30

Shurtleff might have disparaged the best legend of Nix's Mate, but old Bostonians will still tell you that the words shouted by Nix's mate were: "God, show that I am innocent. Let this island sink and prove to these people that I have never stained my hands with human blood."

And, of course, sink it did.

Geologists tell us many of our islands will disappear, even Long Island eventually, but time, the sea, the wind, the ice carved our continental islands, and the factual lore of their beginnings can be observed as we explore them.

The Making of an Island

THE ORIGIN of islands has always fascinated man. By the time of the Crusades, speculations of all kinds were current. Were they formed before or after the deluge? Were they made of fire and ice? Giraldus Cambrensis gave his opinion that sometime after the Flood, islands came into the world, not violently but out of the accretion of alluvial deposits. With exceptions such as the Aeolian Islands, Ceylon, even Trinidad, he was partly right as we can see if we examine deposits of clay, marl, and sand on Martha's Vineyard or Nantucket, Fire Island or Long Island proper, and the islands of the Mediterranean. All of these and many more are continental islands that were once part of the mainland . . . but many have known violence, the violence of ice.

Most of the world's islands are upstarts. At the most they are new summer visitors in the old established colony of the earth. The earth has a distinguished age; the scientists make it into a nice round figure, 4,700,000,000 years. Few islands can compete with that. Martha's Vineyard, for example, with its less than 100,000,000 years, can barely conceal its origins. We can explore it, finding out its past, making it lay bare its secrets, as we might go through an old New England house unraveling the stories. The clay will tell us a great deal and in no place can you find more extraordinary clay than that fabulous range of color found in Gay Head.

ISLANDS

The Canary Islands today. (Courtesy Elizabeth and Charles Wilds)

32

Painter, writer, tourist—yes, even geologist—has found the extraordinary passion of these cliffs unforgettable. As in a child's finger painting made out of the thick mud, here are the colors of the rainbow: a white shining as brilliantly in the sun as a glare of bare beach, then the sudden Matisse contrast of coal black and then lines of green and red, tan, gray, and yellow. If you try to climb down the cliffs you find sand and gravel and clay. You could pick some up and hold it in your fingers and ask it to tell you all, but the smear of color cannot tell the whole truth. On this land there were once camels and horses. On a great plain farther inland sharks once swam in the meadow. The great amphitheater of Greek shape that is called The Devil's Den, made by the digging out of clay for commercial purposes in years gone by, has made the cliff itself look perishable, vulnerable. They used to say that volcanic flames sometimes came from The Devil's Den, and the old Indians told stories that an Indian giant lived there, sitting cross-legged, patiently boiling whales over a fire made of the largest tree. Yes, there were great forests here, and sometimes pieces of ancient tree trunks, lignite, can be found telling Gay Head's history; scientists have revealed well over a hundred different kinds of plants that grew in this rainbow of clay. You could hear the wind through the willows, the pines, and the oaks. You could feel the sharp leaf sting of the holly and the sassafras. You could smell the sweet sequoia and magnolia. And then these ancient forests turned into swamp and the swamp turned to sea. The Cretaceous age had begun and with it the flooding of the earth's surface, the submergence of fifty percent of the earth's surface.

Who was there to watch the last of the great trees die? Who was there to see the mud beat down upon the meadows? Only the clay can tell us and there is still a hundred and fifty feet of Cretaceous clay on the Vineyard. But there was no island then. The true Vineyard was yet to be formed, but it began its true struggle about 95,000,000 years ago.

Eventually the great sea retreated; 70,000,000 years ago the sea level fell, the Cretaceous period turned into the Cenozoic, all life changed. Gone now was the dinosaur; here came the mammal. But nature was quixotic; it still was not satisfied. Some 30,000,000 years ago the sea took over again, with the Vineyard then lying just below the surface in water sixty feet deep. Life grew more familiar. You could find the Vineyard scallop there now, quahogs and cockles, the burrowing razor clam and mussels.

The sea withdrew farther. It would rise again slightly, but the sea's work was over; the summertime of creation had disappeared; it was now time for winter and the ice to take over.

Just the change of a few degrees of temperature started a new cycle 33 of weather. In the high altitudes the snow lingered into spring and into early summer. By midsummer it had not melted but congealed into packed-down snow, crystallized over and over again by melting and refreezing. The snow, once soft and light, turned to something heavy and oppressive. The ice began to move sluggishly. It grew thicker until there were blankets of ice such as there still are over Greenland and the Antarctic. They were thickest, of course, where the snow had been heaviest; over North America, Europe, and Asia, there were areas 8000 feet deep. Then like New England molasses it began to pour and stick to the land, to move slowly southward over the landscape. Pleistocene ice massed together in three areas: over northeastern Quebec and Labrador, over the Keewatin area west of Hudson Bay, and over the Canadian Rockies.

Ice masses developed and merged with one another, overlapping and intermingling. The southern compound ice sheet reached out and began to form one of the great islands of the world—Long Island—and to shape and form the continental mass of Cape Cod, which would further break down into the Elizabeth Islands, Martha's Vineyard, and Nantucket. The ice moved as slowly as a recalcitrant walker on the shores of our islands today. It edged its way a few yards a month—sometimes skidding as a car might on the highway, sometimes moving slowly and

St. Pierre, Martinique, before the earthquake.

sternly like the proverbial tortoise. It moved more rapidly along the valleys, slower in the uplands. As it moved its edge became irregular. Where the topographic depressions were major it moved faster, forming lobes that were advance sentinels for the coming of the main body of ice.

As the glaciers accumulated in shallow areas and valleys frost took its toll, breaking down the rocks, moving loose soil and boulders and scraping further materials from valley walls. As the ice moved it pushed in front of it materials that would be redeposited in the landscape as the ice melted. And as it moved it rasped and scraped the land and the rock ledges; it scoured as might a diligent housewife the bottoms of the valleys; like a bad child it scratched at the rock floors and finally began to form an end or terminal moraine that has left its mark curiously on some of the islands we most love. It created the backbone of Long Island and shaped the Vineyard and Nantucket.

These moraines were the outermost limits of the ice and its jumbled mass of rocks that would finally terminate as sand, forming such superb islands as Fire Island.

Terminal moraines range in size from yards to miles as the ice edge oscillated back and forth during the centuries. Sometimes great chunks separated from the sheet, slumped down and formed hollows, or kettles, such as the kettle holes of Long Island and the Cape. Finally the ice slowly receded, cold weather gave way to milder temperatures, and with the slow, deliberate passage of centuries the sea level changed. During the ice age the sea level was 150–300 feet lower than it is today. The shoreline changed; the New England coast, for example, was probably a hundred miles east of where it stands today. All of the terminal moraines left distinctive marks, none perhaps so much as on the island of Nantucket, where the geologist J. B. Woodworth has said is found "one of the most instructive portions of the terminal moraine of the last ice epoch in North America because it is the most distinct and isolated of the glacial accumulations. Set in the waters of the ocean far to the south of the morainal belt of Cape Cod a distance nearly its own length from the neighboring island of Martha's Vineyard, its peculiarities of its glacial form despite the long relief of the island are readily discerned." But in the hills of the Vineyard, too, we have, as the geologist Frederick Wright

Southern New England as shown on William Wood's seventeenth-century map. The islands of the area were produced by the great ice sheets of centuries past.

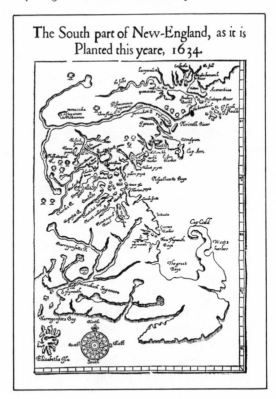

The South part of New-England, as it is Planted this yeare, 1634.

35

ISLANDS

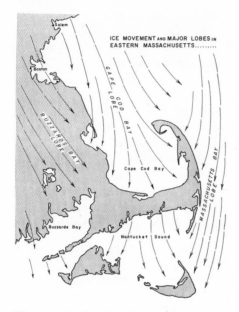

Diagram picturing glacial movement in the Ice Age. The so-called lobes were advance sentinels for the coming of the main body of ice.

described it in 1890, "one of the most remarkable true terminal moraines anywhere to be found in the world.

"Throughout their whole extent these terminal accumulations form a marked feature in the landscape rising for a considerable portion of the distance from a hundred and fifty to three hundred feet above the general level of the country and being dotted over with huge boulders transplanted at a greater or less distance from the north."

Sharing one terminal moraine with Long Island, this trinity of islands formed by ice, sea, and wind are the stories of an island's beginning. And new islands will be formed—perhaps breaking off from the debris of the present land—while some islands may disappear. Long Island, for example, may be an Atlantis of another era, disappearing in the way of some of the smaller islands, like Nix's Mate off Boston Harbor.

For such islands there is always the mark of the sea, and for many of our barrier islands the continuation of their existence is partially dependent upon the stabilization of the plant life that deters erosion.

Fire Island, for example, owes its existence in part to the movement and deposition of sand by long shore currents. The waves, striking the coastline at an angle, set up a current parallel to the shore, transporting

The earthquake wave at St. Thomas. Earthquake waves are frequently misnamed "tidal" waves. Neither winds nor tides cause these waves. They are produced by subterranean convulsions which lift or otherwise agitate the surface of the land or the ocean bottom. (Nineteenth-century print from the authors' collection)

37

Krakatoa during . . .

38 sand in a westerly direction. Storm waves, too, have constantly changed the outline of the island. Inlets have their own story. Today when the oldest and largest community on Fire Island—Ocean Beach—could well become another inlet to Great South Bay (a ten-block-wide section of the village is definitely endangered, according to Gene Gilman, a field officer for the State Department of Conservation), a quick look at Fire Island's vulnerability is well in order.

That the changing positions of the inlets is not just a recent phenomenon may be seen from the older records. Fire Island Inlet is reported to have been opened in the winter of 1690–91, at which time it was nine miles across. About 1754 there are said to have been seven inlets east of Fire Island Inlet, each from one quarter to one half mile wide. In 1797 an inlet into Moriches Bay and one opposite Bellport were open. By 1829 the inlet to Moriches Bay was closed, and the Bellport inlet was filled in about 1834 as a result, according to historian Osborn Shaw, of being blocked by a sunken brig. A map prepared in 1843 for Benjamin F. Thompson's *History of Long Island from its Discovery and Settlement to the Present Time* shows an opening into Shinnecock Bay but no others east of Fire Island Inlet. The great storm waves which now and again wash over the island are thus nothing new to Great South Beach. It is just that more people are now aware of, and affected by, the results because more people are living there.

. . . and after the eruption. (Authors' collection)

The sea is cruel to islands, but to those islands whose very existence 39 depends on their volcanic nature there are other threats. If nature can create an island, it also can destroy it. The very volcanic nature of some islands means that they contain behind their own navels the very source of their destruction. Take a look around any volcanic island and you will find a strange deception in the landscape. True, the verdant, rolling hills are now covered with tropical plants, the deep crevices make superb harbors, and boats languish at the docks. Only the tips of the small mountains have about them an air of agitation as the clouds hover around them—the false smoke of the fire that once raged and erupted into these hills and mountains.

The volcanic island, curiously enough, found as it is throughout the Caribbean and the South Seas, has been the island of romance and poetry, of calm and peace, but always smoldering behind the facade is the possibility of disaster. Islands have often been devastated by earthquakes and volcanic eruptions. In 1089 England was severely shaken. In 1137 Cautania, in Sicily, was destroyed with the loss of 15,000 lives. In 1596 thousands were killed in Japan. In September, 1603, 100,000 people were killed in Sicily. In 1692, 3,000 were killed in Jamaica. In 1703, 200,000 perished in Japan. The list of earlier disasters is overwhelming, but few are remembered with the exception perhaps of the eruption on the island of Krakatoa in 1883.

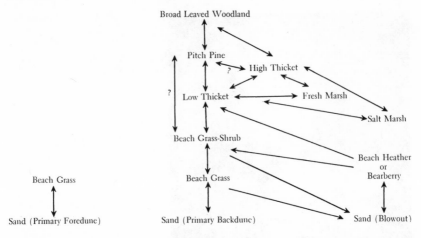

Related types and possible directions of change of plant communities on Fire Island. Upward or leftward arrows indicate change to be expected when factors favoring biotic succession outweigh retarding physical factors. Downward or right-facing arrows indicate change which may occur when unfavorable physical factors are dominant over biotic. The more or less stable zonation across the island is probably a result of a balance between the opposing factors. Uncertain relations are indicated by question mark. (Courtesy U. S. Department of Interior)

40

In modern time no contemporary earthquake so shook the popular mind as the so-called "great island calamity" of 1902—"The Martinique Horror." The contemporary accounts say: "No such frightful calamity, unequaled for the suddenness of the blow, the number of victims, the completeness of the desolation, has ever come upon the civilized world with such overwhelming and howling forces."

It was a calamity that came as do many volcanic eruptions, almost without warning—or what warnings were given were ignored by the population. Mont Pelée, the volcano, had been asleep. It nestled there, a quiet crater, surrounded by villas and attractive resort homes of French West Indians. The contemporary accounts say a week before its great eruption it had given some indication of activity. Those who lived near the mountain had heard strange sounds and at midnight of May 3 the sky was suddenly lighted by a belching forth of boiling mud. Then all was quiet except for a little intermittent noise, but on Ascension Day at ten minutes of eight in the morning the people of the city of St. Pierre heard a devastating explosion. Lava enveloped the city and the shipping; a tidal wave struck the road.

During the Spanish-American War, Martinique had come into the news frequently and so all eyes in the United States turned to the telegraphic reports that came from St. Pierre, now called "the modern Pompeii." One report came from St. Thomas:

The city of St. Pierre, the principal port of the French Island of Martinique, was destroyed, with all its inhabitants, at 8 o'clock on the morning of May 8th by a flow of lava from the volcano Mont Pelée. The number of lives lost was believed to exceed 25,000 and may be as great as 40,000.

The whole top of the mountain was reported to have blown off. For three minutes lava and ashes poured down upon the doomed city. The panic-stricken population fled to the waterside, but in vain. Eighteen ships in the harbor were destroyed by molten lava, and the people who fled to the wharves were soon caught in the awful flood and consumed.

All the suburbs within a radius of four miles were destroyed. Cable communication with the island, as well as with the islands of St. Vincent, Barbados, Grenada, and Trinidad, was interrupted. Steamers that escaped from the vicinity during the eruption reported the losses as follows:

City of St. Pierre and suburbs, with from 25,000 to 40,000 inhabitants. . . .

Sixteen steamers, names and nationalities unknown.

Alligator resting in its natural habitat, Everglades National Park, Florida. (National Park Service photograph by Jack E. Boucher) 41

Map drawn at the time of "The Martinique Horror."

As far as is known only thirty persons were believed to have survived of all those who were at St. Pierre at the time. These were taken by the French cruiser *Suchet* to Fort-de-France. The commander of the cruiser reports that by one o'clock on Thursday the entire town of St. Pierre was wrapped in flames. He endeavored to save about thirty persons more or less burned from vessels in the harbor. His officers went ashore in small boats seeking for survivors, but were unable to penetrate the town. They saw heaps of bodies upon the wharves, and it is believed that not a single person in the town at the time escaped.

The only vessel to escape from the harbor was the British steamer *Roddam*, which arrived at St. Lucia the following day. . . .

Of the eighteen vessels destroyed in the harbor three are said to have been American. . . .

Among the survivors of the *Roraima*, one of the vessels destroyed, was Assistant Purser Charles Thompson, who left us the following vivid account:

We left Antigua at midnight, May 7, and arrived off Martinique at daylight May 8. Our people had seen the fire from Mont Pelée for many miles at sea during the night, and now as we came up into the roadstead in the day light the pillars and waves of flame gushing out of the top of the volcano appeared to be rising a hundred feet in the sky. Several of the passengers came on deck early to watch the eruption of the volcano. . . . The water of the harbor was quite smooth, and although enormous quantities of flame and smoke were boiling up out of the crater of the volcano the sky was not darkened and the view was excellent. The West Indian and Panama Telegraph Company's steamer *Grappler* was lying moored to a huge buoy. Thus she proved a big screen between us and the fury that rolled down upon us afterward.

Not one of us on the *Roraima* would have escaped with his life if the *Grappler* had not been in a position to protect us. . . . I went on deck early and found many of the passengers and all of the crew who were not on duty below, lined up on the port side, watching the show. As it was Ascension Day no one in St. Pierre would do any work. . . . The piers, the streets, and in many cases the housetops, were covered with spectators, enjoying the majestic spectacle.

All of the [*Roraima's*] passengers except Mrs. McAllister, who, on account of her delicate health, remained in her second cabin stateroom on the port side of the main deck amidships, were lined up on the port rail enjoying the sight. Most of the crew were lined up on that side, too. I don't suppose that there were twenty persons below out of our ship's company of sixty-eight. Captain Muggah had not yet left his bed in his room under the bridge.

While Moreley, Brown and I were standing in the alleyway on the starboard side of the ship, not far from my stateroom, which was a little forward of midships, we heard a terrific explosion on Mont Pelée. The sound seemed to crush everything flat. We saw that Mont Pelée had burst open about one-third of the way from the top and fronting us.

There gushed out of this great vent, which was fully a quarter of a mile wide, an awful mass of fire, thousands of times greater in size, but like the gush of fire that darts out [of] a cannon that is fired at night. At the same time the sea began to boil in frothy waves, as if stirred up by some power, the movement far below the surface. In less than a minute the fire leaped from Mont Pelée down upon the city of St. Pierre, struck the water with a frightfully loud hissing sound, and came rolling over and over upon itself as it advanced upon us.

I never saw anything like the rolling of this wave. It advanced like a gigantic beach comber of flame, with its top part always rolling down and under the mass, and with the after part of it constantly rising up to a height of more than a hundred feet as the great mass of flame leaped in our direction. Vast clouds of steam arose from the contact of fire and water. The fire seemed to blot out the city of St. Pierre from our sight. There rose up

44   Aerial view of the Ten Thousand Islands. (Courtesy Florida Development Commission)

Lush foliage and giant ferns are characteristic of the Hawaii Volcanoes National Park. (Courtesy National Park Service)

Islands have often supported unusual fauna. Within our own country the Key deer of Florida is such an example. A miniature deer, it is now protected on Great Pine Key. (Courtesy United States Fish and Wildlife Service)

an outcry of myriads of voices. Now, as the fire wave advanced close to us, the steam arose in great clouds and cut off from view what was left of the ruined city. . . .

At that moment the *Roraima* was tossed about in the boiling sea and a great whirlpool pulled her far over on the port side. Then the terrible hurricane of fire struck her and heeled her far over on the starboard side, so that she lay almost on her beam ends. At the moment this fire wave swept over us I heard a noise, frightfully loud and threatening. That was the sound of our two masts and the smokestack and the port side of the bridge being swept away like chaff. Even then we had not received the full force of the fire blast, for the cable steamship *Grappler* served as a screen for us. I was told that later the *Grappler* was flung down on her side, blazed up in flame in every part of her hull and plunged down beneath the water all in an instant.

I ran into my stateroom on the starboard side of the ship; my idea was to plunge under the bedclothes in my berth and so protect myself from the wave of fire and gas from the volcano, but before I got half way covered the fire hurricane hurled the ship over almost on her beam ends on the starboard side. The porthole of the stateroom was wide open, and the green water came dashing in in a great force. It was almost scalding hot.

The inrush of the water swept me off my berth, and I staggered out into the middle of my room. The water was so hot that I felt as if I was burning up, and I madly tore off my coat and waistcoat. As the ship rolled still deeper to the starboard I thought she was going to turn turtle and sink. As the *Roraima* lay wallowing in the sea and trying to pick herself up I held fast to the electric light fixture. I jumped outside then, into the gangway. Captain Muggah was on the bridge giving orders and trying to save his ship when the blast of flame overwhelmed him. . . .

Things were no better on the British island of St. Vincent. There the volcano Soufrière, quiet since 1812, rose up some four thousand feet above the island. For three weeks it had been grumbling like an angry old man, but still the islanders paid little attention and by the seventh of May the crater was pouring out mud and vapor, clouds were rolling above the volcano eight miles and higher; yet some could stand there fascinated by the cloud shapes, looking like heads of cabbage, they said, or beautiful flowers, the vapors extending for miles, making sheets of electric fire. The natives said that the mountain cried "Vo-vo-vo-vo." Soufrière was a mud volcano, and ash and sand and rock poured into Georgetown. They say that the first stones were cold as ice, but when they hit the ground they split open, spilling a heart of hot coal. This countryside of thatched houses burned instantly. The hot ashes dug into every corner; great hot rivers deterred the traveler. The thirst was ghastly; those who lived could remember only that terrible thirst.

The last of the Caribs, the Indians descended from those who had welcomed Columbus, died on the slopes of Soufrière.

In addition to the islands made of fire and those made of ice there are a small group of islands that during the nineteenth century and continuing well into the twentieth century were a matter of great dispute as to their origin. These, of course, are the coral islands and the accompanying reefs. "Everyone must be struck with astonishment," wrote Charles Darwin, "when he first beholds one of these vast rings of coral rock often many leagues in diameter, here and there surmounted by a low verdant island with dazzling white shores, bathed on the outside by the foaming breakers of the ocean, and on the inside surrounding a calm expanse of water."

For years the old voyagers talked about them with different names, some calling them lagoon islands, others atolls, barriers, encircling reefs, fringing reefs, or shore reefs. But they had all been fascinated by the beauty, serenity, and curious makeup of these islands.

Naturally it was Captain James Cook who first brought back information on the Pacific coral reefs. He described such an underwater reef on his trip to Palmerston Island:

At one part of the reef, which looks into or bounds the lake that is within, there was a large bed of coral almost even with the surface, which afforded one of the most enchanting prospects that nature has anywhere produced. Its base was fixed to the shore, but it seemed to be suspended in the water, which deepened so suddenly that at a distance of a few yards there might be 7 or 8 fathoms. The sea was at this time quite unruffled, and the sun shining bright, exposed the coral in the most beautiful order— some parts branching into the water with great luxuriance, others lying collected in round balls, and in various other figures—all greatly heightened by the spangles of the richest colors that flowed from a number of large clams that were everywhere interspersed.

It was that great scientist Charles Darwin who, voyaging in mind as well as body on the British vessel *Beagle*, in 1835 brought back the most information about coral reefs. Darwin felt that these reefs and coral islands had been subsiding volcanic mountains on which the upgrowing reef attached itself the way a jewel might to a rock. And indeed jewels these extraordinary corals are, great submarine jewels that are some of the most beautiful phenomena in all our natural world. No matter how the coral reefs started (although Darwin's theory—after much dispute— is generally accepted), its real owner, its proud sculptor is of course the

The Alae eruption in 1963, Hawaii Volcanoes National Park, Hawaii. (National Park Service photograph by Garrett A. Smathers)

framework of colonies of calcareous animals and plants—with coral the main support like the good father of a family.

It was not until the underwater world became available to us with skin and scuba diving that the beauty of these corals made us alert to the greatest undersea gardens of the world. Coral itself is closely related to that little beauty the sea anemone. The coral is a plankton feeder paralyzing the food with its stinging cells. Just what these stinging cells can do is all to familiar to those of us who have been burned by the malevolent fire coral in tropical waters. The individual coral animal builds itself a protective skeleton. As with the clam, this skeleton is an exoskeleton, the material that protects it as strong as limestone—calcium carbonate. They are gregarious these corals, forming themselves into colonies of from hundreds to millions, populating tropical waters with all the frenetic passion of urban builders. But they are drifters too, drifting in ocean currents and re-establishing themselves in other suburbs of the sea.

In the United States with all our other wonders we have one extraordinary formation of islands that actually include dead coral reefs. They are, of course, the Florida Keys.

View of Haleakala Crater from the observatory at Haleakala National Park on Maui Island, Hawaii. (National Park Service photograph by Cecil W. Stoughton)

We used to go to the Keys as children, boating from one island to another when the form of communication was still as ancient as that used by the Polynesians who first landed on the atolls of the Pacific. Then the typical Florida catboat was always manned by some old captain who knew the channels of the Keys the way he knew conch chowder and the bite of chiggers. Each boat carried a conch shell, not for display on a coffee table, but because it was used to cry its call from island to island as we would dock.

Now we approach the islands differently over that great transverse highway and elevated ribbon of cement that does seem curiously appropriate because the islands themselves have some of the cement of the earth beneath them. The great highway seems almost like another coral structure. stretching out into infinity, the glare of the Florida daylight burning the eyes, the trip becoming almost perilous at times in the blinding sunshine. Below are the islands, but the traveler catches only a glimpse. He must still get off and putter around, sampling the local delicacies, a beautiful monotony of fried fish and shrimp.

But even now the mystery of the Keys can haunt one: Where have they come from? What are they? How do they exist? In a way it seems like a land of the dead past, a land of bone and desolation—that curious desolation that one can get only surrounded by all the tropical beauty, the rich green of the mangrove, the blatant blue of the water and the silence of corroded rock. Offshore the coral reefs guard the island chain. Some of the Keys themselves are simply dead reefs that must have existed thousands of years ago, but this land of dead rock was formed by the living and one gets some of the feeling of that extraordinary vitality in the passion of the sun.

Coral reefs and islands usually occur only on the eastern shores of continents. You will never find them for example in California nor on the Pacific coast of Mexico. But they are abundant in the West Indies, on the coast of Brazil, the East African coast in the tropical areas, the northeastern shore of Australia where the Great Barrier Reef runs for well over 1,000 miles, and our superb coral coast of Florida—200 miles of treasured islands to those who love them.

The Keys are divided into two different types of groups. There are the Eastern Islands making an arc from Sands to Logger Head Key—long narrow islands studded with low shrubs at home on the coral limestone. They move into the shallow waters of Florida Bay, dipping their feet into water almost too hot for swimming upon occasion and supporting mangrove swamps.

50

The Pine Islands, or the Western Group, are of a different nature. They have limestone rock shaped by the sea. The Keys are ever changing; a new polyp will form, a new coral structure take shape; a new mangrove seed will deposit itself and perhaps start another islet. Sometimes the islands form their own bridges of mangroves; there is always some new island or islet being formed by coral animals or mangrove seedlings.

The Keys are new islands compared to the other islands of our country; their ages can be measured in tens of thousands of years. Once this area was a plateau; beyond its edge the water was warm enough for the coral to take root. The sea level dropped and the coral luxuriated; then came disaster. At one point in the late Wisconsin glacial stage the ocean fell too low for the reef to live. It had to start all over again, but you'll find traces of the old reef even now. Go underwater from one of the Keys and you will see the reef lying a distance away. Beyond the reef, because really both Key and reef are hilltop, the water descends abruptly 100 . . . 300 . . . 500 fathoms. Throughout all this area there is a variety of coral with beautiful sea fans and sea whips.

At night this reef is even more mysterious. Now the reef moves with live animals, the coral thrusting out tentacles. While we watch these fascinating reefs we wonder why they occur just in the warm waters.

Probably it is because the calcium carbonate of the animal life is soluble in cold water rather than in warm. Sometimes strays will appear in other places with tropical conditions, in Bermuda for example; but although reef-building corals normally need such water, corals are found in other waters: those of Alaska, the fjords of Norway, the New England coast, the Cape of Good Hope. The false coral occurs in the Mediterranean, and another species in Japan.

Florida has many superb islands and within the Keys, as we have said, there are mangrove islands, but on the southwest coast there is a superb collection of mangrove islands that are one of the natural-history wonders of the world. They are called the Ten Thousand Islands.

Mangrove islands, of course, are found in many parts of the world from the Persian Gulf to the Gulf of Mexico, in Central America and the atolls of Indonesia, small islands in the Amazon as well as the Malay Archipelago. The mangrove tree makes its own world. It is a perpetual child, its feet in water, forming sometimes an impenetrable, prickly jungle that lashes the skin. It is not a shell that makes these islands, nor the glaciers, nor the waves themselves, but the simple mangrove seed, the most peripatetic of all travelers. Certainly the red mangrove of Florida has traveled a great distance, perhaps even from the West African coast, idling along on the westward current. Mangroves need the rain of the tropics, their aerial roots being too high to always get sufficient water. If you look at any of the old maps of the Ten Thousand Islands, the area was simply marked "dangerous" and most sailors, except old hunters and fishermen, bypassed it. The islands are still relatively undiscovered and untouched, strangled in their strange geography and matted with the red mangrove. 51

Sometimes on these islands you will find the white and black mangrove hovering behind the red mangrove. The black mangrove used to be called "salt trees" when we were young, and many times we have been at fish fries with old fishermen who would put one or two of the leaves of the black mangrove in a good pot of conch stew. Their leaves always have some salt on them and it is a good sauce for the fisherman's larder. Sometimes as children we were sent to pick some of these leaves. It was a perilous task indeed; their breathing roots, or pneumatophores, stand up like small sentinels on the ground. They prick the feet as viciously as the spines of a cactus. Behind the black mangrove there would then be a few white mangroves looking very frequently like buttonwood trees. But the great island builder is neither the black nor the white but that fine, almost deep chestnut-colored tree called the red mangrove.

There is something strangely compelling about the mangrove, so compelling that it interested Alexander the Great enough to make him halt his march through Asia simply to observe mangroves making land. Columbus also observed them in the New World and hoped they indicated that the area of the Caribbean was peaceful. How otherwise, he said, could plants exist in water? Certainly it meant hardly any wind and an almost Eden-like area.

As children we used to go with an old Cracker, poling our way through the Ten Thousand Islands, looking for a place to moor. Looking through the branches, you could see nothing. Each island appeared a jungle and indeed each one was. Then you could look down into the water, far different from the crystal gin color of the Keys. Here the water was as black as Irish tea. Below, we were told, were turtles, sponges, fish of all kinds and beautiful coral, but you could not see them. The water was opaque from the tannin in the bark of the mangrove. Often we would see floating on the water a seed barely visible, shaped roughly like a small Italian cigar, making its way on some journey to start some new island—a seed stubborn and persistent, that would eventually hit a sandbar and give birth to an island.

These islands struck us as mysterious—almost holy in their rich solitude—and it is not surprising that throughout the world islands have indeed been holy places, places of pilgrimage, repositories of ancient civilizations. The same creative and religious instinct—the quality of awe—that allowed the Indian of long ago to carve the deer mask on Key Marco in Florida haunted those pilgrims in whose steps we now follow to the holy islands of Lake Titicaca.

52

3

Holy Islands

THEY MOVED slowly, a procession of dark shadows carrying in their hands lighted flares. These were the pilgrims to the sacred islands in Lake Titicaca, that superb lake fastened like some breast shield between Peru and Bolivia. Their way was always slow and tedious, given to long fasts and the propitiation of the gods.

Along the way there were stone statues that proved that the gods themselves could turn you into stone if you did not follow the proper laws as one followed this holy path. On they went until they reached the lake itself. Sudden panoramic views caught their eyes beyond the ruddy bulk of the sacred islands of Titicaca. On the shore of the lake they had to rest and wait for the balsa boats which are still used on the water.

This great lake, a hundred and twenty miles long, sixty miles wide, is 12,488 feet above the sea. Although the lake is deep and its height from ground level amazing, it nonetheless gives a kind of warmth to the islands on it. But the wind is wicked—the prevailing winds coming from the northeast—and with the narrow balsa boats navigation is difficult. The balsa boat is nothing but a raft made up of a bundle of reeds tied together. The pilgrim poised on his knees in the middle while a companion like some Charon out of ancient myth poled him to the sacred island.

Iona, the Scottish island of St. Columba's ancient churches. (Nineteenth-century print from the authors' collection)

54   Lake Titicaca is filled with these holy islands: Amantene, Taqueli, Soto, Titicaca, Coati, Campanario, Toquare, and Apouto. The largest is Titicaca and it is the most sacred of all. It was here that the Incas were created according to their early stories; here the children of the sun started out on a journey to bring religion and art to the savage tribes before the Incas settled on the mainland. The legends of Manco Capac and his wife-sister Mamma Occlo sprang up in these islands: Manco Capac carrying his golden rod, moving northward to the spot where he would finally sink it into the ground and fix the seat of the great civilizations of the Incas. Then he moved on to the spot where Cuzco now stands, and founded there the City of the Sun, the capital of the Inca empire.

It is the sacred islands—the island of Titicaca dedicated to the sun and the island of Coati dedicated to the moon—that perhaps are the most romantic examples of the many holy islands throughout the world. The ancient civilization worked hard to make the islands sacred. First, laborers had to remove the rocks, then craftsmen had to build elaborate terraces which they covered with earth ferried by the balsa boats from the mainland. The resultant temples and island terraces rivaled the pyramids of the Egyptians.

Next they brought corn, cultivated it and held it sacred. It grew so poorly that each grain was holy, each grain would be sent as a gift from

This huge statue of Buddha located in the suburbs of Changhua, central Taiwan, is one of the largest images of Buddha in the Orient. The seventy-two-foot statue is made of brick and concrete. (Courtesy Pacific Area Travel Association)

Boaters drift around the Itsukushima Shrine on Miyajima, Hiroshima Prefecture, Japan. (Courtesy Japan National Tourist Organization)

56

Brightly caparisoned elephants in the religious procession held annually in Kandy, Ceylon, to pay homage to the Sacred Tooth Relic of the Buddha. (Courtesy Ceylon Tourist Board)

heaven to the Inca temples that were springing up over the countryside. If in some way an Indian was able to get a grain of this holy maize, grown on this sacred island, it was said he would never lack for bread throughout his life.

And here, too, are the remains of that great Temple of the Sun, one of the superb buildings of ancient times. The pilgrim, however, did not come ashore and make his way directly to the temple. There were all sorts of preliminary fasts, purifications and penitences to be handled. He had to go through portals—the Door of the Puma where a priest of the Sun listened to elaborate confessions, then on to the Portal of the Bird, finally to the Gate of Hope itself, and on to the Sacred Rock. This is the most sacred rock in all of Peru. Here no bird alights, no animal ever comes near, no person may put a foot. Here the sun was created to illuminate the world, and the ancients plated this rock with gold and silver to imitate the sun and the moon. Today, of course, it is nothing but a weather-worn mass. Beyond it is the labyrinth where passages once led to rooms that were perhaps for the Indian type of Vestal Virgins.

The gold offerings described, indeed all the gold of the Incas and the silver mines of the South American coasts, led innumerable explorers to decimate the peoples and culture. Halfway around the world, in Ireland, there was no gold—only the coin of the spirit to be constantly reevaluated on the holy islands of that country.

If the Latin-American Indians and later colonials with their elaborate churches were people of great ritual, the old Irish hermits who often sought an island to contemplate or to write out those old books of penance or to transcribe ancient documents for seats of learning throughout the world had far less elaborate temples or even dwelling places.

The Irish, of course, have always been obsessed with the idea of islands. They have written stories of elaborate adventures and voyages where a visit from one island to another almost follows the progress of a soul itself. In the *Voyage of Maildun,* for example, there are the island of the monstrous ants, the terraced isle of birds, the island of the wonderful apple tree, the island of bloodthirsty quadrupeds, the isle of red-hot animals, the island that died, the island of the burning river, the island of weeping, the isle of the four precious walls, the isle of speaking birds, the isle of the big blacksmith, an island guarded by a wall of water, an island standing on one pillar, the isle of intoxicating wine fruits, the isle of the mystic lake, the isle of laughter, and the isle of the blessed; and then there is one story about the hermit of the searock. These searocks of Ireland are tremendously emotional experiences visited by sailing over a torrential body of water in the native boat, a small curragh of

Ireland. The legend of the hermit of the searock, for example, is about a man born and bred on the island of Tory off the coast of Donegal where there was a monastery dedicated to St. Columcille. The hermit of the searock of the Maildun story threw himself against the bare rocks and prayed incessantly. And he told this story:

I was born and bred in the island of Tory. When I grew up to be a man, I was cook to the brotherhood of the monastery; and a wicked cook I was; for every day I sold part of the food entrusted to me, and secretly bought many choice and rare things with the money. Worse even than this I did; I made secret passages underground into the church and into the houses belonging to it, and I stole from time to time great quantities of golden vestments, book-covers adorned with brass and gold, and other holy and precious things.

I soon became very rich and had my rooms filled with costly couches, with clothes of every color, both linen and woolen, with brazen pitchers and caldrons, and with brooches and armlets of gold. Nothing was wanting in my house, of furniture and ornament, that a person in a high rank of life might be expected to have; and I became very proud and overbearing.

One day I was sent to dig a grave for the body of a rustic that had been brought from the mainland to be buried on the island. I went and fixed on a spot in the little graveyard; but as soon as I had set to work, I heard a voice speaking down deep in the earth beneath my feet: "Do not dig this grave!"

I paused for a moment, startled; but, recovering myself, I gave no further heed to the mysterious words, and again I began to dig. The moment I did so, I heard the same voice even more plainly than before: "Do not dig this grave! I am a devout and holy person, and my body is lean and light; do not put the heavy, pampered body of that sinner down upon me."

But I answered, in the excess of my pride and obstinacy, "I will certainly dig this grave; and I will bury this body down on you!"

"If you put that body down on me, the flesh will fall off your bones, and you will die, and be sent to the infernal pit at the end of three days; and, moreover, the body will not remain where you put it."

"What will you give me," I asked, "if I do not bury the corpse on you?"

"Everlasting life in heaven," replied the voice.

"How do you know this; and how am I to be sure of it?" I inquired.

And the voice answered me, "The grave you are digging is clay. Observe now whether it will remain so, and then you will know the truth of what I tell you. And you will see that what I say will come to pass, and that you cannot bury that man on me, even if you should try to do so."

These words were scarce ended when the grave was turned into a mass of white sand before my face. And when I saw this, I took the body away, and buried it elsewhere.

It happened some time after that I got a new curragh made, with the hides painted red all over; and I went to sea in it. As I sailed by the shores and islands, I was so pleased with the view of the land and sea from my curragh that I resolved to live altogether in it for some time; and I took on board all my treasures—silver cups, gold bracelets, and ornamented drinking-horns, and everything else from the largest to the smallest article.

I enjoyed myself for a time while the air was clear and the sea calm and smooth. But one day the winds suddenly arose and a storm burst upon me, which carried me out to sea, so that I quite lost sight of land, and I knew not in what direction the curragh was drifting. After a time, the wind abated to a gentle gale, the sea became smooth, and the curragh sailed on as before, with a quiet, pleasant movement.

But suddenly, though the breeze continued to blow, I thought I could perceive that the curragh ceased moving, and, standing up to find out the cause, I saw with great surprise an old man, not far off, sitting on a crest of a wave.

He spoke to me; and, as soon as I heard his voice, I knew it at once, but I could not at the moment call to mind where I had heard it before. And I became greatly troubled, and began to tremble, I knew not why.

"Whither art thou going?" he asked.

"I know not," I replied; "but this I know, I am pleased with the smooth gentle motion of my curragh over the waves."

"You would not be pleased," replied the old man, "if you could see the troops that are at this moment around you."

"What troops do you speak of?" I asked. And he answered: "All the space round about you, as far as your view reaches over the sea, and upwards to the clouds, is one great towering mass of demons, on account of your avarice, your thefts, your pride, and your other crimes and vices." He then asked, "Do you know why your curragh has stopped?"

I answered, "No," and he said, "It has been stopped by me; and it will never move from that spot till you promise me to do what I shall ask of you."

I replied that perhaps it was not in my power to grant his demand.

"It is in your power," he answered; "and if you refuse me, the torments of hell shall be your doom." He then came close to the curragh, and, laying his hands on me, he made me swear to do what he demanded.

"What I ask is this," he said; "that you throw into the sea this moment all the ill-gotten treasures you have in the curragh."

This grieved me very much, and I replied, "It is a pity that all these costly things should be lost."

To which he answered, "They will not go to loss; a person will be sent to take charge of them. Now do as I say."

So, greatly against my wishes, I threw all the beautiful precious articles overboard, keeping only a small wooden cup to drink from.

"You will now continue your voyage," he said; "and the first solid ground your curragh reaches, there you are to stay."

60

The island of Malta, where St. Paul was shipwrecked.

He then gave me seven cakes and a cup of watery whey as food for my voyage; after which the curragh moved on, and I soon lost sight of him. And now I all at once recollected that the old man's voice was the same as the voice that I had heard come from the ground when I was about to dig the grave for the body of the rustic. I was so astonished and troubled at this discovery, and so disturbed at the loss of all my wealth, that I threw aside my oars, and gave myself up altogether to the winds and currents, not caring whither I went; and for a long time I was tossed about on the waves, I knew not in what direction.

At last it seemed to me that my curragh ceased to move; but I was not sure about it, for I could see no sign of land. Mindful, however, of what the old man had told me, that I was to stay wherever my curragh stopped, I looked round more carefully; and at last I saw, very near me, a small rock level with the surface, over which the waves were gently laughing and tumbling. I stepped on the rock; and the moment I did so, the waves seemed to spring back and the rock rose high over the level of the water, while the curragh drifted by and quickly disappeared, so that I never saw it after. This rock has been my abode from that time to the present day.

For the first seven years, I lived on the seven cakes and the cup of whey given me by the man who had sent me to the rock. At the end of that time the cakes were all gone; and for three days I fasted, with nothing but the whey to wet my mouth. Late in the evening of the third day, an otter brought me a salmon out of the sea; but though I suffered much from hunger, I could not bring myself to eat the fish raw, and it was washed back again into the waves.

61

I remained without food for three days longer; and in the afternoon of the third day, the otter returned with the salmon. And I saw another otter bring firewood; and when he had piled it up on the rock, he blew on it with his breath till it took fire and lighted up. And then I broiled the salmon and ate till I had satisfied my hunger.

The otter continued to bring me a salmon every day, and in this manner I lived for seven years longer. The rock also grew larger and larger daily, till it became the size you now see it. At the end of seven years, the otter ceased to bring me my salmon, and I fasted for three days. But at the end of the third day, I was sent half a cake of fine wheaten flour and a slice of fish; and on the same day my cup of watery whey fell into the sea, and a cup of the same size, filled with good ale, was placed on the rock for me.

And so I have lived, praying and doing penance for my sins to this hour. Each day my drinking vessel is filled with ale, and I am sent half a wheatflour cake and a slice of fish; and neither rain nor wind, nor heat, nor cold is allowed to molest me on this rock.

Such rocks or Skelligs exist around Ireland, none more exciting than the Great Skelligs off Dingle Bay. These are three rocky islets, the most southwest extension of Eire; the Great Skellig rises up nine miles to the southwest of Port Magee and appears but a black splotch on the ocean. In the old days they used to warn you that you should never attempt a trip to the Skelligs without taking at least four men in the curragh because of the great feat of navigation that it required. It is a trip past many islands out to the Great Skellig. Some have thought these islands lead to the island of the blessed. One passes Puffin Island before one reaches Lemon Rock, then moves on to Little Skellig and finally to Great Skellig. Great Skellig was a monastery every bit as exciting as those ancient Peruvian and Bolivian temples. It was nature itself that carved out the cathedral here, not man. The old books explained:

The rock towers higher and higher and splits into fantasic forms like the open leaves of a book set upright with strips of bright green running between them or fringing the horizontal blocks of the strata at their feet. When the sunlit mists or vapors sweep in driving clouds above them, the effect is in the highest degree mysterious and beautiful; but when at one moment these mists rise so as to entirely conceal the heights, and next they vanish as if at the touch of some unseen hand, and the cliff again stands revealed against the blue unfathomed sky, it seems as if the whole scene were called up by some strange magician's wand.

62   The ancient approach to the monastery from the landing place was on the northeast side. There are six hundred and twenty steps from a point in the cliff which is about one hundred and twenty feet above the level of the sea up to the monastery. The rest of this flight of steps is broken away and a new approach was cut in very recent times. The old stairs ran in a varying line; the steps which grow broader toward the upper half of the ascent are alive with cuffs and long cushions of the seapink and at each turn the ocean is seen in foam hundreds of feet below.

For centuries this was an isle of pilgrimage with a rock halfway up the side known as Christ's Saddle. Sometimes the rocks would be carved into rude crosses. No one knows completely the true story of the Skelligs, but throughout the ancient writings there are brief references. Some say the Milesians were wrecked off this coast and that the great prince, St. Milesius, was buried on this island. St. Patrick knew it, perhaps, but certainly since the days of St. Patrick it has been an island of "piety and devotion." It was an island of birds and was originally named according to the old writings for the Archangel St. Michael, the same St. Michael to whom Mont-St.-Michel in France and Michael's mount in Cornwall, England, are dedicated.

Ireland, too, is studded with crannogs. These were ancient Irish lake dwellings generally placed on an island. Often the island has disappeared and only the crannog will remain like some large mound in a bog.

63

Initial letters as drawn in the Book of Kells. (From *Celtic Illuminative Art,* by the Rev. Stanford F. H. Robinson; published by Hodges & Figgis & Co., Ltd.)

There is no more famous holy island than Iona, the final spot of pilgrimage for St. Columba as he moved eastward to Christianize Scotland and England. It was St. Columba, the greatest of all the itinerant saints of Ireland who in the year 563, his biographers tell us, went for the love of Christ with twelve disciples to Britain. He was looking for an island where he could practice the Christianity that he was about to teach. He found his island. Some call it the most beautiful island in all of the Western Isles of Scotland. It was an ancient island once used by the Druids and its old Gaelic name was Innis Nan Druinich—The Isle of Druidic Hermits. But in Columba's time it began to be spelled Ioua, perhaps an old Norse term, and eventually it became Iona. Just about three and a half miles long and a mile wide, it has rugged moors on the north and south, a fertile central belt, and is almost the center of the archipelago of islands in the west. To the north are the isle of Rum and the low shores of Canna; distantly lies Skye; looking west are isles of Coll and Tiree. South one looks over to Mull. Certainly Columba was at

64

This is the Inca Temple of the Sun on the Island of the Sun on Lake Titicaca in Peru-Bolivia. (Courtesy Grace Line)

the center of a small universe; here was a place for a central monastery. Columba, who is credited with writing many poems, wrote, many scholars feel, the following poem on Iona shortly after he landed. It expresses superbly the love of nature, the attraction of a bookish life, and the spirit of devotion that inspired the island saints of the time:

> Delightful would it be to me to be in Uchd Ailiun
>   On the pinnacle of a rock,
> That I might often see
>   The face of the ocean;
> That I might see its heaving waves
>   Over the wide ocean,
> When they chant music to their Father
>   Upon the world's course;
> That I might see its level sparkling strand,
>   It would be no cause of sorrow;
> That I might hear the song of the wonderful birds,
>   Source of happiness;
> That I might hear the thunder of the crowding waves
>   Upon the rocks;

That I might hear the roar by the side of the church
    Of the surrounding sea;
That I might see its noble flocks
    Over the watery ocean;
That I might see the sea monsters,
    The greatest of all wonders;
That I might see its ebb and flood
    In their career;
That my mystical name might be, I say,
    Cul ri Erin (Back turned to Ireland);
That contrition might come upon my heart
    Upon looking at her;
That I might bewail my evils all,
    Though it were difficult to compute them;
That I might bless the Lord
    Who conserves all,
Heaven with its countless bright orders,
    Land, strand, and flood;
That I might search the books all,
    That would be good for any soul;

65

Inca well on the Island of the Sun on Lake Titicaca. (Courtesy Grace Line)

At times kneeling to beloved heaven;
    At times at psalm-singing;
At times contemplating the King of Heaven,
    Holy the chief;
At times at work without compulsion;
    This would be delightful.
At times plucking dulse from the rocks;
    At times at fishing;
At times giving food to the poor;
    At times in a carcair (solitary cell).
The best advice in the presence of God
    To me has been vouchsafed.
The King, whose servant I am, will not let
    Anything deceive me.

They started building, Columba and his faithful followers, making their monastery out of wood and wattles, making the small mill and a kiln. The guestchamber itself was wattled, the cells for the monks were constricted. Columba himself occupied a small cell, with a stone pillow for a headrest, built in the high part of the ground. None of these buildings remain, of course.

66   The monks had taken a solemn monastic vow and they were called by Columba his *Familia* or chosen monks. There were also working brothers who would attend to the cattle and adolescent pupils. It was not unusual for some young man to be sent as a student to Columba, whose monastery become known in faraway Europe.

Across the water was Scotland itself, dominated by the king of the Picts. It would be two years before Columba would try to convert the powerful Pictish nation. The old Scottish history of Columba says that he made this conversion by first making the sign of the cross over the Pict Royal House. The door opened immediately. Again he made the sign of the cross over the head of the king, and the latter dropped his sword from a withered hand. "And so it remained until he believed in God, and being made faithful to God, his hand was restored."

Twelve years later St. Columba had founded monasteries on most of the surrounding islands.

Eventually these islands would know the Vikings, but the Vikings also had their own holy islands—one for example the beautiful island of Gotland, of which the old sagas say: "In the days of old a fair and beautiful island, low and dim floated on the sea by night and the people beheld it as they sailed to and fro; but each morning at sunrise it disappeared beneath the waves until the evening twilight would come again when it would rise and float over the surface of the Baltic as before. No one dared to land upon it though the belief was general that it would become fixed if a fire were lighted there." Finally Thjelvar did land there, lighted a fire, and the island became stationary; to this day the bay is called Thjelvarvik.

England, too, has had its holy islands, none more beautiful than Lindisfarne, the eastern isle of the saints. Some call it Iona's daughter, but this island is reached differently than Iona, which must be reached by boat. One is dependent upon the tide to get access to Lindisfarne, just as one is with that holy island of Mont-St.-Michel.

All of these islands have been at one time or another islands of pilgrimage, islands which, as Sir Walter Raleigh said, could give one a "scallop shell of quiet."

> Give me my scallop shell of quiet,
> My staff of faith to walk upon,
> My script of joy, immortal diet,
> My bottle of salvation,
> My gown of glory (hope's true gage)
> And then I'll take my pilgrimage.

67

4

Oh, Manhattan!

In the old days they made a careful difference between Manhattan Island and the Island of Manhattan. The old friends that used to gather in the quilting bees of the old churches of New York, less than fifty years ago, even then carefully differentiated the Island of Manhattan from Manhattan Island, a small river island that had disappeared years before.

It had been important enough at one time, however, for James Fenimore Cooper to explain in *The Spy*:

> Every Manhattanese knows the difference between "Manhattan Island" and "the Island of Manhattan!" The first is applied to a small district in the vicinity of Corlaer's Hook while the last embraces the whole island; or the city and county of New York, as it is termed in the laws.

Corlaer's Hook jutted into the East River in the vicinity of Rivington Street, and there rested Manhattan Island—barely more than an acre, surrounded by salt marsh and dependent upon the tides, it was often covered with more than a little seawater.

To the ever-changing landscape or rather cityscape of the Island of Manhattan, the octogenarians of fifty years ago made a major contribution. They told stories that were sometimes legends, legends that were sometimes truths. They drew a picture of Manhattan that made it truly an island, a rural one, of old homesteads, of neighborhoods in

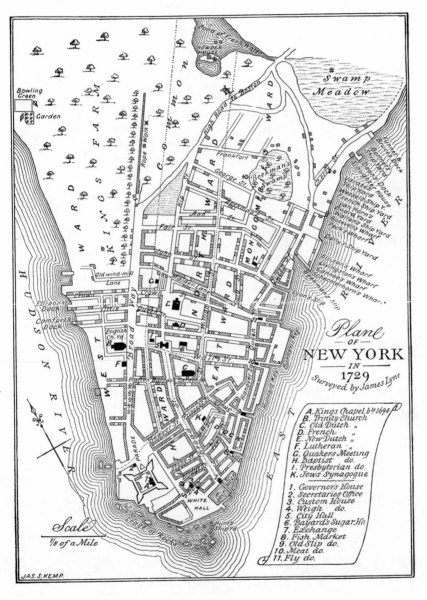

Plane
OF
NEW YORK
IN
1729
Surveyed by James Lyne

A. Kings Chapel, b.t 1694
B. Trinity Church
C. Old Dutch "
D. French   "
E. New Dutch "
F. Lutheran "
G. Quakers Meeting
H. Baptist   do.
I. Presbyterian do.
K. Jews Synagogue

1. Governor's House
2. Secretaries Office
3. Custom House
4. Weigh   do.
5. City Hall
6. Bayard's Sugar Ho.
7. Exchange
8. Fish Market
9. Old Slip do.
10. Meat do.
11. Fly do.

Scale
⅛ of a Mile

JAS. S. KEMP.

Old Jefferson Market.

A bird's-eye view of Governors Island. (Nineteenth-century print from the authors' collection)

which everyone knew everyone else, of sudden passions of wisteria on old wooden buildings, of old tales about someone who remembered someone who once lived in a house in Greenwich Village when there was nothing between the Village and the East River, "a mite over the way."

Everybody walked in those days, and they knew their Island of Manhattan better. Even now perhaps the island has more walkers than any comparable territory and on the superb tours of the New York Historical Society, or in the quiet wandering from one street to another in old neighborhoods, it is not too difficult to conjure up some of the apparitions of the past.

As children we used to be entranced by those who remembered—or said they did—Barnum's great museum facing St. Paul's Church at the corner of Anne Street. It was far better, they assured us, than any other museum. It had in it, for example, the club which killed Captain Cook, a picture of the nurse who had held George Washington; it had strange mermaids and even stranger mastodons. There were those who remembered the old-time theaters, the minstrels of that great spot, Niblo's Garden, where the superb pantomimes entranced young and old. In those days if there was a ballet at Niblo's Garden even the gentle- 71 men decently covered their eyes with their hats. Overdressed, overheated, the dancers whirled away in a world of their own, every bit as naughty for the times as the ungarbed on today's off-Broadway stage.

There were those who remembered the trees of New York, not the pathetic ones we have today (Jane Jacobs points out that there are twenty species we could still satisfactorily grow in our streets), but the great superb shade trees, everything from Stuyvesant's pear tree to the flowering trees at Canal Street, where wild flowers bloomed in the spring and young children fished in the water.

All these, of course, were memories, even then. No one was old enough to really remember that early New York, but there was a time when there was still a folklore of the city that was passed down from one generation to another, so that one felt that one had actually seen George Washington on that corner, had brushed the arm of Lafayette, had walked through Lispenard Meadows.

One old historian said, "I do not think that people can understand the size of our city in those days. We all knew 'who was who.' Old Mrs. Stewart, in black brocade selling candy by the penny's worth at Chambers and Greenwich Streets; Katie Ferguson on Hudson Street, making all the jelly and sweetmeats and Mrs. Isaac Sayres in Harrison Street preparing all the wedding cakes were types of the time. Everybody knew them all, knew the ministers and our few rich men."

ISLANDS

If there were those who knew the rich and mighty, who whipped down the road on coaching days, there were others who knew the excitement of strange dark corners—the Bowery, where the Boston stage used to arrive; that strange and awful corner called Seven Points, the most evil corner in the world. They had stories, too, of crime—the harbor police constantly catching smugglers on the river, the gangs that roamed the streets.

Some spots seemed to have more color than others, or indeed, in the oppressive days of New York, simply more of a breeze. One such place was the Battery Park, where one could watch steamers coming from distant spots. The Battery still has a little of that old feeling—really old New York, really Dutch New York.

Then the tide still went across the beach; there were orioles and bluebirds in quantity. Castle Garden still existed, and in the area around Bowling Green and the Battery were the homes of the great. Johnny Battin, an old gentleman, once talked with that wonderful cityscape historian John Flavel Mines, who wrote under the name Mr. Felix Oldboy for the *New York Evening Post* and the *New York Commercial Advertiser*, from 1886 to 1890:

72

"Felix," said Johnny Battin, "I like to come here to the Battery and think of all the changes I have seen hereabouts in the last seventy years. Yes, it was seventy years ago since I saw the British flag hoisted on the battery that stood back there by the Bowling Green. We camped up in East Broadway the night General Putnam evacuated these barracks and stole up along the Hudson to Fort Washington. That night a terrible fire broke out by the riverside here, and swept up Broadway, carrying away Trinity Church and nearly every other building as far as St. Paul's. It was a terrible conflagration, and lit up everything almost as clear as day. The houses were nearly all of wood, and by daybreak more than a third of the city was in ashes. The brick houses on Broadway, opposite the Bowling Green, were all that were left standing, and there Lord Howe made his headquarters. . . .

"It looks like a long way over to Staten Island, but I remember when the bay was frozen over solid from the Battery to what is now the Quarantine grounds. Our troops crossed over on the ice from Staten Island, and dragged their cannon with them. I carried the orders from Lord Howe, and it startled them, I can tell you; but they came through all right. . . .

"I have seen a great many changes, my boy, in seventy long years, and I am more than ninety-four. But it has been a pleasant and happy life, and its happiest part has been lived in my little home on Greenwich Street. It won't be long now before I am called to meet the King of Kings, but you will live to see greater changes than any I have known. Love your country, boy, and love your home. It's an old man's advice and the secret of happiness."

The low, circular architecture of Madison Square Garden Center provides a startling contrast to the soaring spires of the Empire State Building and surrounding towers of Manhattan. (Courtesy New York State Department of Commerce)

New York City in the age of sail. (From oil painting by Charles Robert Patterson; Courtesy The Mariners Museum, Newport News, Virginia)

74

Wollman Memorial Skating Rink, New York City's Central Park. (Courtesy Pan American Airways)

Walt Whitman, born an islander, had a passion for New York ferries:

Living in Brooklyn or New York city from this time forward, my life, then, and still more the following years, was curiously identified with Fulton ferry, already becoming the greatest of its sort in the world for general importance, volume, variety, rapidity, and picturesqueness. Almost daily, later, ('50 to '60) I cross'd on the boats, often up in the pilot-houses where I could get a full sweep, absorbing shows, accompaniments, surroundings. What oceanic currents, eddies, underneath—the great tides of humanity also, with ever-shifting movements. Indeed, I have always had a passion for ferries; to me they afford inimitable, streaming, never-failing, living poems. The river and bay scenery, all about New York island, any time of a fine day—the hurrying, splashing sea-tides—the changing panorama of steamers, all sizes, often a string of big ones outward bound to distant ports—the myriads of white-sail'd schooners, sloops, skiffs, and the marvellously beautiful yachts—the majestic sound boats as they rounded the Battery and came along towards 5, afternoon, eastward bound—the prospect off towards Staten Island, or down the Narrows, or the other way

The Stuyvesant Pear Tree.

75

The Gracie Mansion.

The Dakota Apartment House, New York City, in 1882.

up the Hudson—what refreshment of spirit such sights and experiences gave me years ago (and many a time since.) My old pilot friends, the Balsirs, Johnny Cole, Ira Smith, William White, and my young ferry friend, Tom Gere—how well I remember them all.

There were those who remembered the cries of old New York, cries still in existence during World War I, shouted by street sellers in horses and wagons, sometimes keened by a woman selling horseradish. One old woman, the corn lady at Hester and Bond Streets always sang:

> "Hot corn, hot corn;
> Some for a penny and some two cents.
> Corn cost money and fire expense,
> Here's your lily white hot corn."

They sold everything on the streets in those days. Pears and molasses, oysters, pure spring water for two cents a pail, the latter coming from the great springs of Greenwich Village.

The City Dock, too, collected the passerby. He could look out upon schooners from the West Indies, the activity of commerce on South Street—soon to be restored. The old City Dock itself, the one built originally by the West India Company, had known the exports of the New Netherlands: the herbs and the lumber, the grain. Then moving on in time, it was visited by brigs of the great traders with the West Indies, a "pink" from Barbados, a full-rigged bark from London, a slave trader from the coast of Africa. Riches poured in from the rest of the world, from faraway islands, silks from China, cashmere from India, sandlewood and perfume, and from Ceylon, gum and spice.

Along the shores of the island they would look out on other islands, those delightful islands and rocks of the East River, that glorious spot where Hell Gate is a maelstrom worthy of Edgar Allan Poe. And the names they had: Patrock and Frying Pan—terrible to sailors—Floodrock, Hogs Back and Ward's Island—then a lovely island surrounded by a rich hoard of lobsters and embowered with trees. Felix Oldboy said:

Poetry has yet to discover the rare beauties of the East River, whose water-front is not surpassed in attractiveness in any country. Gemmed with islands, garlanded with woods, beset by rocks which are rich in legendary lore, and headlands that are redolent of history; in many spots as unchanged as in the days when Harlem was a tiny, sleepy settlement, remote from the busy City of New Amsterdam; this arm of the sea is one of the loveliest, if least regarded, features of the grandest of American cities.

When New York was created to be a great maritime city, care was taken to supply it with all that it should need in the way of islands, and they were strewn about its main island foundation with proper pic-

76

turesqueness. Those who remember the islands in their primeval loveliness, when they were the homes of some of our ancient families, and were clad in verdure in summer, and in impressive dreariness in winter, may regret that the city has been compelled to use some of them as homes for the sick and the sinner, but even the stern majesty of the law cannot make them other than beautiful. It is a matter of congratulation with those who believe that the useful need not be ugly, that there are some things which the hands of men who fancy that they can always improve upon nature cannot mar. The islands in the East River will always remain an enchanting feature in the topography of this maritime metropolis, and New Yorkers, who are somewhat prone to overlook advantages which lie directly at their doors, will some day open their eyes to appreciate them.

Those were the days when Governors Island was sometimes called Nut Island—the name originally given to it by the Indians—where chestnut trees grew in abandon and even after the time of the Civil War private fishing rights were given to old Long Islanders in need. Then there were Little Barn Island—now called Ward's—and Randall Island and Blackwell's Island, once known as Long Island when Long Island was still Nassau. There was Dead Man's Rock, where Captain Kidd was supposed to have buried treasure.

Old historians recreate even the Revolution, the old magazine at Turtle Bay, the bands of privateers sweeping across the channel at Hell Gate to capture the guard of the Bay. As late as 1814 it was rumored that New York was going to be invaded by a British army. And the old fortifications lasted in memory of the boys and girls who had climbed over them. Some remembered poor broken-down cottages in Central Park, while others knew the great houses, occasionally one still standing—Gracie Mansion, for example, where Washington Irving was a frequent guest in the house of Archibald Gracie, "an old gentleman with the soul of a prince." Now, of course, it is the home of the mayor of New York.

The island itself had islands within the islands, stretches of country or cityscape that were units unto themselves—none more exciting to most New Yorkers than that island within: Greenwich Village, an area that we are able to document and where perhaps, like Leopold Bloom, we can make a day's pilgrimage, reconstructing it through nineteenth-century documents and newspapers.

The Village streets are conducive to roaming. This has been true since most of them were country lanes. Take any Sunday in the Village and you will find whole families on foot, examining streets that they know well, looking into windows, contemplating the seasonal changes of Washington Square, commenting on the seasonal misfortunes of the

77

Art Show, investigating the local efforts to keep a kind of rural atmosphere in the Village, those wonderful, vibrant, aggressive turkeys in the window of Comullo's meat store, the spasmodic lights decorating all the shops for the Christmas season. In the Village these things are done a little bit better than in other places; there's a touch of poetry to the blaringness, a touch of whimsy to the salesmanship.

The Village, it appears at times, is a Village of shopkeepers, and to prove that it has been a thriving community since the latter part of the nineteenth century, let's roam awhile as the Villagers do.

Perhaps the most instructive walk would be one undertaken by some contemporary Leopold Bloom. You will recall, in those Gargantuan pages of *Ulysses*, Bloom started out on foot to collect the advertisements for the newspaper he represented. In so doing, he crossed and recrossed Dublin on foot so often that the very fabric of the city came alive. One can do the same thing by taking a page out of Joyce's wonderful phantasmagoric map of time and space and following an advertising solicitor a hundred years ago in the Village. Oh, yes, Virginia, there was advertising then, and the Village shopkeepers in their ads were extraordinarily proud of their stability and honesty. So then roam through the Village, first, as Leopold Bloom did, starting with a breakfast of pork kidney, obtainable from some magnificent butcher stall around Jefferson Market. That imaginary advertising solicitor is a gourmet; as he eats his pork kidney, he ponders on the different paths he will take during the day.

It is a day in late fall. The sun is burning unseasonably on the sidewalks, but our advertising man decides to start his day by the river. The sun burns the top of the river with some warm affection that the Village is blessed with in October. So down goes our wanderer to the foot of West 13th Street. There a magnificent sign rivaling the Palisades signs of today—only this time on our side of the Hudson—obstructs the view of the river. It says "Charles L. Buci." Mr. Buci is the greatest importer of yellow pine. It comes from faraway Jacksonville in Florida. The Buci Company is famous particularly for the great ship the *Louis Buci*, which has made one of the fastest trips up the coast with a glorious load of yellow pine.

(The waterfront still attracts Villagers. They climb over all sorts of obstructions to reach the beckoning fingers of the docks. You can hear a concert or two upon them now, or dream of the distant summer in the fall as you watch the sunlight on the water and absorb the magnificent mystery of the port. A hundred years ago the port extended far up here to the Village, so that there was an extraordinary amount of activity going on down at the waterfront. Beyond Mr. Buci's sign you could see coastal· steamers, sailing ships, ferries, the boat to Albany all competing for the majesty of the waterway.)

78

The famous oyster family of Mersereau—I. P. Mersereau from Staten Island—are selling oysters at the foot of Perry Street. Moses Corson, on No. 17, is selling oysters and clams at the foot of 10th Street. Mr. Pillsworth is selling oysters and clams aboard barge 24 in the river. There is a great largess about Mr. Pillsworth. Oysters and clams are not his only outlet for creativity; he designed some of the famous yachts that you will see in the Hudson—the *Atlanta*, the *Greyling*, and the *Montauk*.

One of the high points of the day is yet to come. Our man is enough of a boy to appreciate Tuesday in the Village. He looks at some of the round-faced clocks and knows he has just enough time to get up there. He must be on East 13th Street by 11 o'clock. As he approaches it the street is mobbed with men and young boys. The cries "Oh, he's a fine one, he's a fine one! What do we get, what do we get?" start reaching his ears from 5th Avenue. Suddenly he stops. For a moment there the stable smell is almost overwhelming. This is the Village Horse Fair. It goes on twice a week—every Tuesday and Friday—at 11 o'clock, and here are the largest horse auctioneers in the city. Nearly 10,000 horses change hands each year, and well over 15,000 carriages and wagons. Occasionally the old baymen from Long Island will come in selling marsh grass—because that's used extensively throughout the stables of the Village—for bedding down horses. There are some real entrepreneurs who are even importing peat from Ireland and Scotland, but it will never really take the place of that wonderful island marsh grass properly dried for the proper horses of the Village. Any horse fair is a heady experience, and our man is caught up in the pleasure of the horse flesh itself, the movement and excitement of the street, the fantastic cries of the auctioneer, the neighing of the horses, the brittle sound of old carriages on the cobblestones, the fierce competitive atmosphere of the fair itself.

There is a hawker going up and down selling raw oysters, and our man, who has already sampled the wares of each barge, tries one more; this one, of course, could be the one to turn the stomach. He will skip lunch. He has sampled too many oysters, smelled too many horses, had too much excitement thus far in his day. He needs a change of scene.

He goes over to 15th Street. He, like Leopold Bloom, is quite a man with the ladies and on 14th Street one will find the greatest theatrical center of the world. First he will see Mr. Rodriguez, the greatest importer of Havana and Key West cigars, at 62 E. 14th. He will sit for a while on those fashionable bent chairs of Mr. Kohn's on 14th Street, proudly advertised, says Mr. Kohn, "because he has purchased an entire forest from Europe."

Then he will reach his destination, the heart of the theatrical suppliers. There is Koehler, one of the most famous costumers, at 2 Union

79

A rainy ride in another city island—Hong Kong. (Courtesy Pacific Area Travel Association)

Square. He will supply anything from armor to children's clothes. There is Mr. Winkleman's, where Lillie Langtry goes to get her wigs. Our solicitor hopes she may be there today, but she isn't. That bright sun has suddenly disappeared and our New York Bloom decides he needs a sweet. He goes over to Percy Rockwell's at 3rd Avenue and 9th Street, one of the largest caterers and confectioners in the city. You can see their fifteen bakery wagons all over town but particularly in the Village. He obtains a little sustenance and an ad there, and makes his way uptown again, crisscrossing as only the Villager does, back and forth, east and west, north and south, many times during the day.

The day grows colder and colder. He skirts the very south end of the Village, picking up some sample flowers at John Meehan's, a manufacturer of flowers, ostrich and fancy feathers, and particularly an inventor of the bellflower, on Houston Street. Just around Mr. Meehan's factory you will find the little work-sick children whose parents are in home industry making those very same flowers. But our solicitor cannot let con-

Battery Park (extreme right) marks the southern tip of the Island of Manhattan. At the top of the picture East River separates the island from Brooklyn. The two bridges are the Williamsburg and Brooklyn Bridges. (Courtesy New York State Department of Commerce)

science rule him. What must move him is feet—walking, walking, walking over the Village. On 14th Street, he finds Mr. Side, a fine advertiser, a great promoter of the period. In his window is one of the largest Bengal tigers ever caught. He is of course a furrier, one of the greatest in the Village, and the furriers will eventually subsidize one of the new arts that will emerge in the Village—the motion picture.

Finally, his day over, our man finds himself outside of that thriving center, Jefferson Market. In the old days morning started early around Jefferson Market, but for small children it was a delightful place to go. You were very fortunate if one of the maids decided to take you along to wander in and out of the sawdust shops, to pick up something from the barrels, to get a delightful "good morning" from all who worked there. The market itself was sealed in. The middle door was the one by which you entered this great world of cold and cabbage. Out of the corner of your eye you could see all the game of the day, having been shipped from Great South Bay on the Island where the old bay folk seemed to make a living out of supplying these game birds for the wealthy tables around Washington Square. All this, however, was captured, as childhood so frequently is, out of the corner of the eye. Mabel Osgood Wright, the daughter of the famous Unitarian minister of New York, always remembered some of those experiences as she entered the middle door of Jefferson Market as a small child. First there was the pigman. He always greeted her with affection, and he looked alarmingly like what he sold. His color, she said, was the color of the meat of rosy ham, and his hair the clear yellow of the rind of fat. The pigman was naturally a pork butcher, but like all pork butchers of the day, he also sold butter, eggs, and cheese.

Most of the people of the Market in those day were English. The pigman's mother wore one of those half wigs called frisettes and the proverbial tartan shawl, tossed, in the way of the mongers, across the chest and tied in the back. Her gown was always black—spotlessly black if black can be spotless—which meant it shone as though it had been treated with a high patent-leather polish that was further accentuated by the cleanest of white aprons across the front.

In those days people shopped in respectable clothes and the shopkeepers were equally respectable. There wasn't any dashing off to the corner in shorts. The pigman's mother would have been shocked by such a lack of propriety. It would have never entered her head—it could not, because her head was so well guarded. Beginning at her shoulders was her fabulous headgear that was almost impossible to describe. It was something between a sunbonnet and a half of a coal scuttle, said Mabel Wright. It was decorated of course with lace, the lace making her appear as though she were an overdecorated Labrador retriever. These were the same caps one found in the middle of the nineteenth century in any of the small provincial towns in England. It was silly to tie these elaborate bonnets' strings and instead they were rolled up and pinned into delicate decorative rosettes beneath the ears.

From the pigman one moved on'across the aisle to a haven of vegetables bursting with the year's vigor, and poultry hanging from hooks on an iron grill. Tyson was the man who sold the poultry, and next to him was DeVoe, who, some said, practiced short weight but always denied it and made it up if you mentioned it. If you complained, you would finally find that you were getting more than your share, at least one more chop in the pound. Far down the aisle was Davis, the fishmonger, the most exciting stall outside of the pigman's. The children used to play in the seaweed that harbored the oysters and clams that came out from Long Island, and there they could find, still alive a sand dollar perhaps, a fiddler crab, even a starfish.

82

You could be in this romantic world of Jefferson Market and then suddenly be scared right straight down to your patent-leather boots if the great fire bell began to ring. It tolled the very torments of the Village. After all, the Village in those days was still a wooden wonder and firemen's disaster. Every man, woman, and child would tremble within the Market itself. You counted the strokes and thus found out where the fire would be. The only way you really knew for sure was by carrying your little pink book. This was a book sold at the candy store called "Taffy John's" on Greenwich Avenue at 11th Street. There is still a candy store there today selling penny candy as of old, but they do not sell that invaluable little pink book which located the fires of the past. You listened to how many times the bell rang and then consulted the book in your pocket. More than one would rush out of Jefferson Market and chase the fire engines. It was easy enough to do. They were all drawn by horses, and the volunteers simply ran a little behind the horses straight to where the flames would now be gathering with terrible speed.

Some of the big fires of course drew absolutely everyone. When Barnum's Museum—the one located at Vandam Street and Broadway—burned down, everybody in the Village went; those that didn't, such as sick children, were promised that "the fire was such a grand one it won't go out today and you can see it tomorrow. There will be smoke and smolder enough, I'll warrant ye, and we'll have a grand time seeing what isn't there anymore."

In and around the Market one found areas of ill repute. At night-time 6th Avenue became a haven for pads—those quiet-moving burglars—for ladies who had concerns other than buying at the Market, and for the flotsam and jetsam that might make their way over from Five Points to the excitement of the court.

There was no doubt that one lived on an island, however. The rivers and their perils were omnipresent.

In the summer of 1878 William O'Neill, otherwise known as Nan the Lifesaver and Nan the Newsboy, set up with two other young men—none of them were over twenty—the Volunteer Life Saving Corps. They went on duty at 7 A.M. and continued until half past ten or eleven at night. Within six months they had saved twenty-five lives in the river. The policemen of the area were extremely grateful for their work, but some said that those whom they rescued often gave them the "blackest ingratitude."

And being an island, it had its pirates. The river pirates had a large rowboat with bags and tarpaulins and always worked with at least three or four people. They knew how to muffle their oars so successfully that you could never hear them coming along the wharves. They went under

83

the ramshackle piers with all the mystery of Dickens so rapidly and quietly that no policeman could reach them. During the day they had generally singled out the vessel that they would rob; between midnight and morning they went about their business.

At best they were small-scale pirates, however. Old Manhattan had its piracy, but for the real treasures, the gold of storybook lore, Manhattanites would have had to sail, as we must, to other islands.

Arrest of Captain Kidd.

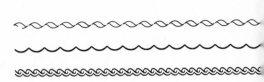

5

Pirate Islands

HE WAS BORN in New England in the farthest village of the Eastern settlement (Maine) on February 2, 1651. His name was William Phips, one of twenty-six children, twenty-one of them sons. He would die His Excellency Sir William Phips, late Captain-General and Governor and Chief of the Province of Massachusetts Bay, New England. But except for the dryness of those vital statistics—even Cotton Mather, who was his contemporary and biographer, could not enlarge upon them—his life was one of color—of drive, ambition, and above all treasure seeking.

Cotton Mather, that staunch old Puritan who suggested that our rewards were in Heaven, rarely on earth, was still enough attracted to him as a person to extol America's greatest treasure hunter.

Certainly the coast of Maine where he was born offered no such treasure to even the most ambitious of young men. He kept sheep in the wilderness, Cotton Mather tells us, and then remarks benignly, "God took him from the sheepfolds." The truth is that William Phips, obviously discouraged by the roughness of the Maine wilderness, became apprenticed in a shipyard located where Bath now is. He was superb with his hands and in four years had learned what he wanted. It was time to move on, this time to that thriving metropolis, that great seaport of Boston. He arrived with enormous self-confidence. He taught himself to

Howth Harbor and Ireland's Eye.

read and write, but most of his interests centered on the shipyards—and a young widow, Mary Spencer Hull.

The marriage was a romantic one and even Cotton Mather had to take pride in the young man's aims. He would be, he told his bride, despite his poverty, the owner "of a fair brick house in the green lane of north Boston." One of the first American businessmen, he, said Mather, cut rather like a hatchet than like a razor; certainly he knew the talent of a hatchet well. He was sometimes harsh in temperament and would upon occasion cut down those who stood in his way. In the beginning nothing stood in his way; even Boston was not big enough for him. He would take himself to the sea. That was the place for "a true temper for doing of great things."

Phips had kept his ears open down on the wharves and had heard from the sailors that came regularly to the docks fabulous stories of Spanish wrecks in the Bahamas. In South America was the golden empire of Spain, linked to the mother country by great ships, galleons with bellies as big as pregnant women's, carrying gold and riches of all kinds. They called these galleons the Plata Flota, the Silver Fleet. They made frequent departures from South America, prey to the terrible storms of the Caribbean, prey to the coral reefs and treacherous currents. It was inevitable that the area would be studded with wrecks and it was inevitable that William Phips would start to search for them.

He began to coast the eastern shore to the Antilles with the usual cargo of cod and wood from New England replaced by the sugar and rum of the southern islands. There were already buccaneers accosting such vessels, and it soon occurred to Phips that he needed a great ship. The money he needed was not available even in Boston. He must go to London and he did so in 1682, making a fabulous sail in his little coastal ship.

He planned, like Dick Whittington, to meet the king, but Dick Whittington had his cat; Captain Phips had nothing but his vessel. How could he even go to court to plead his case with the shabby clothes of a Boston captain? He took a chance as he was always likely to do, sold his ship, ordered fine wigs, velvet clothes and within a year stood before Charles II.

Inevitably the poor Maine boy became a partner with the King of England. He would get a frigate of his own with eighteen guns, the *Algier Rose*, which had once been the property of the Algerian pirates, but had been nicely removed from them. He would get, too, ninety-five men to sail her.

In January, 1684, he set sail for the treasure islands. Once there he showed his administrative ability by organizing his expedition further. With his own hands he started to make a canoe, a giant canoe that would carry eight or ten oars carved from the cotton tree. He slept night after night in the woods, Mather tells us, preparing this "Periaga." They would use his canoe to investigate more closely the reefs and shoals. Mather tells us:

87

This Periaga, with the Tender, being Anchored at a place Convenient, the Periaga kept Busking to and again, but could only discover a Reef of Rising Shoals thereabouts, called, The Boilers, which Rising to be within Two or Three Foot of the Surface of the Sea, were yet so steep, that a Ship striking on them, would immediately sink down, who could say, how many Fathom into the Ocean?

Here they could get no other Pay for their long peeping among the Boilers, but only such as caused them to think upon returning to their Captain with the bad News of their total Disappointment.

Nevertheless, as they were upon the Return, one of the Men looking over the side of the Periaga, into the calm Water, he spied a Sea Feather, growing, as he judged, out of a Rock; whereupon they had one of their Indians to Dive and fetch this Feather, that they might however carry home something with them, and make, at least, as fair a Triumph as Caligula's.

The Diver bringing up the Feather, brought therewithal a surprising Story, That he perceived a Number of Great Guns in the Watry World where he had found his Feather; the Report of which Great Guns exceedingly astonished the whole Company; and at once turned their Despondencies for their ill success into Assurances, that they had now lit upon the true Spot of Ground which they had been looking for; and they were further confirmed in these Assurances, when upon further Diving, the Indian fetcht up a Sow, as they stil'd it, or a Lump of Silver, worth perhaps Two or Three Hundred Pounds. . . .

They had this one further piece of Remarkable Prosperity that whereas if they had first fallen upon that part of the Spanish Wreck, where the Pieces of Eight had been stowed in Bags among the Ballast, they had seen a more laborious, and less enriching time of it: Now, most happily, they first fell upon that Room in the Wreck where the Bullion had been stored up; and they so prospered in this New Fishery, that in a little while they had, without the loss of any Man's Life, brought up Thirty Two Tuns of Silver; for it was now come to measuring of Silver by Tuns. . . .

Thus did there once again come into the Light of the Sun, a Treasure which had been half an Hundred Years groaning under the Waters: And in this time there was grown upon the Plate a Crust like Limestone, to the thickness of several Inches; which Crust being broken open by Irons contrived for that purpose, they knockt out whole Bushels of rusty Pieces of Eight which were grown thereinto. Besides that incredible Treasure of Plate in various Forms, thus fetch'd up, from Seven or Eight Fathom under Water, there were vast Riches of Gold, and Pearls, and Jewels, which they also lit upon; and indeed, for a more Comprehensive Invoice, I must but summarily say, All that a Spanish Frigot uses to be enricht withal.

Mather was not always an accurate historian and we cannot be sure that he has not telescoped three of the trips that Phips made hunting for treasure. We do know Phips had discovered certainly "the greatest treasure of them all," had been knighted by the king, returned home to build that fine brick house in Green Lane, was appointed Governor of Massachusetts, and had other great plans for treasure hunting. They were not, however, to come to fruition and after a visit to London, where he was planning another treasure hunt, he died of "malignant fever," with that dream still unrealized.

The islands of the world have always attracted treasure hunters and pirates, frequently the two going together, but indeed our entire coastal area has its own stories of treasure. Every bit as much of a gentleman as Sir William Phips was Captain Kidd, and his treasure is almost omnipresent, or at least legends of it, on our islands. We do know that he did deposit gold on Gardiner's Island, but he supposedly buried much money, too, on Coney Island, Nantucket, the marshes behind Boston, the Isle of Shoals, Ocean Beach, the Bahamas, the Florida Keys, Money Island, and, of course, Fire Island. Eventually these spots became crowded with legends. On Charles Island, for example, in Connecticut, near Milford, the townspeople discovered that when they dug for Kidd's

88

89

Cruise ship entering San Juan Harbor, Puerto Rico, passes historic sixteenth-century fortress of El Morro. (Courtesy Puerto Rico Information Service)

money a headless man jumped out of an iron chest that they had dug up, hurled himself at them, then leaped into another pit surrounded by blue flames. On Monhegan Island the legend went that actual spirits protected the treasure, and it was necessary if one hunted there in one of the caves to be utterly silent. At the sound of the human voice demons spirit away island treasures. Until recently this legend still lived on in the memory of some of the old Monhegan Islanders and the old axiom "Dig six feet and you will find iron; dig six more and you will find money" would still be quoted to you.

On Damariscove Island, Kidd, so they say, tossed his treasure overboard into a bottomless lake and on Appledore in the Isles of Shoals he buried another lost hoard. For years that island was haunted by the ghost of old Babb—a spectral visitor with a red ring about his neck and phosphorescence encircling his body. Misery Island near Salem also attracted hunters for Kidd's treasure, but none was ever located.

Captain Kidd, being a gentleman of Manhattan, naturally knew the islands around that shore and he supposedly buried treasure on the north side of Liberty Island in New York Harbor. One Sergeant Gibbs tried to dig it up in 1830 with the help, we might add, of a fortune-teller. The fortune-teller, however, did not tell them what else they were likely to find. They did indeed find a box four feet long, but it contained a monster with wings, horns, and blue breath. Gibbs was so shocked that he fell into the water and was nearly drowned.

As you sail your boat around the Island of Manhattan, those enchanting islands all have their stories. It is rumored that skeletons have protected treasure on Hen and Chickens for many years and that if you pass by there and see blue lights those are the sockets of their skulls. But not only Kidd and Bluebeard and the like had treasure; there were other names more unfamiliar: Lord Abercrombie for example, whose treasure is still undiscovered today on Tea Island in Lake George, and Jones's treasure in Fort Neck, Long Island. A century or two ago the rumors of that treasure were strong on the island and indeed it was obvious that Jones must have had some relationship to the demons of treasure, because when he died such a demon in the shape of a crow flew in through his bedroom window. Like Poe's raven, it looked at him until his eyesight failed and then as the death rattle choked his life the crow flew away cawing with triumph. The bird flew directly through the walls of the house, it was said, and the house was eventually abandoned because the crow's exit could never be repaired. Replaced over and over again, the cement fell out each time and the gaping hole remained.

The islands of Carolina, Blackbeard territory, have innumerable treasure tales—and the ghost of a lady pirate, Anne Bonny, born in

90

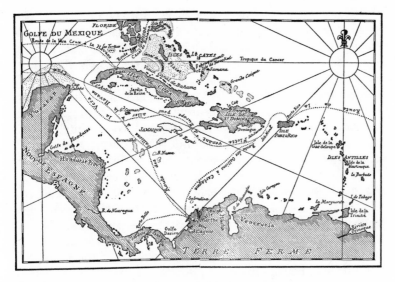

The route of the Spanish galleons in the seventeenth century.

91

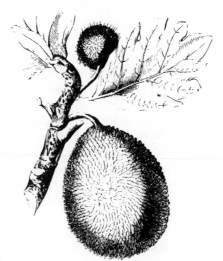

"The Bread-fruit (as we call it) grows on a large Tree, as big and high as our largest Apple-Trees. It hath a spreading Head full of Branches, and dark Leaves. The Fruit grows on the Boughs like Apples. . . . When the Fruit is ripe, it is yellow and soft; and the taste is sweet and pleasant. The Natives of this Island use it for Bread. . . . There is neither Seed nor Stone in the inside, but all is of a pure substance like Bread. . . . This Fruit lasts in season eight Months in the Year; during which time the Natives eat no other sort of Food of Bread-kind." (From *A New Voyage Round the World* by William Dampier)

Ireland. It was Ireland that was to give us two lady pirates, "lady" perhaps being the wrong word in both instances. Anne Bonny was to leave a legend in Charleston in the United States as well as throughout the Caribbean. The other, and the greatest queen of the pirates of all time, was Grace O'Malley of Mayo.

Anne Bonny was born in Cork. Despite the fact that Cork was even then at the dawn of the eighteenth century a relatively thriving seaport, it was, as many areas of Ireland still are, a clannish community. Two of the leading families were the Cormacs and the Sweeneys, both tracing their genealogy back to the mythological past. William Cormac married a charming lady of the Sweeney family. It was one of those ideal marriages, but it was not made in Heaven; it was made in Cork, where, the story goes, there were other beautiful women, some of them forced to become serving maids. One of them—Peg Brennan—was introduced into the Cormac household. William Cormac, an important barrister in the town, was overcome by the gay beauty and wildness of the young girl called Peg Brennan. Perhaps his wife's respectability began to pall, but in any case, Peg became the Peg of his heart and, within due course, his bed. It was not unusual in the eighteenth century for the serving girl to become pregnant by the master, but William Cormac's reaction was somewhat different. His infatuation seemed to be permanent. He could not bring himself to dispatch her to a cottage, as was the custom, but found himself longing to live with her. When the local "nosy parkers" tracked down the affair, Cormac managed to get himself into a battle with the Sweeneys. A riot then occurred that shook the city of Cork, the Sweeneys against the Cormacs, overshadowing whatever cries Peg might have made as she gave birth to a daughter, Anne, on March 8, 1700. The Sweeneys were victorious in the streets of Cork. William, smuggled to England, waited there until he could send for Peg Brennan and Anne so that they might take off for the New World and start another life in another fabulous seaport, Charles Town, now Charleston, South Carolina.

There the strange solitary child who had been born in one seaport discovered another, and there, just as William Phips had been enchanted by the stories of treasures and the free life of the buccaneers (although in his case he considered himself only a proper businessman), Anne began to dream of a less cautious existence. Her father admittedly had contacts with many pirates; after all, were they not the businessmen of the day? And the local Charleston pirate, one Capt. Paul Raynor, a guest often at the Cormac home, dressed like a cavalier.

Escorted often by Captain Raynor, she met on one of her trips to the docks James Bonny, seaman, late of the Royal Navy, in the crew of Conn Kesby, not so much a pirate as a smuggler.

There followed in rapid order a colorful tale of violence, rejection, and escape. Turned against by her father, disinherited, supposedly causing the death of her mother, Anne Bonny at the age of sixteen deserted Charleston for the pirate island of New Providence in the Bahamas. In those first decades of the eighteenth century, New Providence had a criminal population of 3,000 pirates and smugglers. Then the property of the Lord Proprietors of Carolina, the island was a black hole under a blazing tropical sun. Disease was rampant; lawlessness festered in the streets as ripe as a guava. One scoundrel took the place of another and yet curiously enough the records, some disposed of by the Spaniards, and others perhaps disposed of by the more respectable families whose shadows still haunt Nassau today, are unusually weak, but the names of all the great pirates centered at one point on Nassau, then called Charlestown. It was Woodes Rogers, a reformed pirate, who was finally sent to suppress the buccaneer anarchy of Nassau. When Rogers appeared he was greeted by an honor guard—buccaneers who gave him the traditional salute of continuing musket fire. By that time there were those who admitted that they had sinned and would forsake the way of piracy, but there were others who would not. The garden of what is now the Colonial Hotel, for example, bore a fine tree that Rogers used as a gallows and in short time he had dispatched nine who were less willing to go the way of the good.

93

Soon after Anne Bonny arrived in Charlestown, she abandoned Bonny and from there on her attachments were those, said the contemporary documents, of a mere whore, but a delightful one, everyone had to admit. One of the few scoundrels that did not attract her was the violent Edward Teach, otherwise Blackbeard. Anne always seemed to be attracted to the more delicate type—gentlemen pirates and the like— none more exciting nor handsome than Calico Jack—John Rackam. Together (but most of the activity seemed to be on Anne's part) they planned daring exploits; then again like a woman, Anne managed to ingratiate herself by some of her spywork in locating and evaluating the strength of the Spanish troops and warships that had been sent to Cuba, thus earning the thanks of Woodes Rogers himself.

But Anne's career as a pirate was to be tempestuous and short. Eventually her name became linked with the extraordinary Mary Read. The latter, having masqueraded as a boy (she served as a soldier and a sailor), turned pirate. That gentle historian of piracy, Frank Stockton, said, "But although Mary was a daring pirate, she was also a woman, and she fell in love. A very pleasant and agreeable sailor was taken prisoner by the crew of her ship, and Mary concluded that she would take him as her portion of the spoils. Consequently, at the first port they touched she became again a woman and married him, and as they had no other present method of livelihood he remained with her on her ship. Mary and her husband had no real love for a pirate's life, and they determined to give it up as soon as possible, but the chance to do so did not arrive. Mary had a very high regard for her new husband, who was a quiet, amiable man, and not at all suited to his present life, and as he had become a pirate for the love of her, she did everything she could to make life easy for him. She even went so far as to fight a duel in his place, [killing] one of the crew [who had] insulted him."

Events, however, for these two often lovesick lady pirates moved rapidly. Captain Rackam was captured and condemned to death and shortly thereafter the ship on which Mary and Anne were sailing was also captured and they were taken prisoner. Mary was reprieved because she was ill, Anne, or so go the reports, because she was pregnant. Obviously she had been in love again; and some reports would say Michael Radcliffe, her lover, reformed her. In any case having been, she said, "through a spiritual trial of blood and violence," she was now reformed, but Rogers, too, came to her aid. "Was she not repentant?" he said; "should she not be freed?" Evidently so freed, she and Michael took off for the interior of America and her name is lost.

94

If Anne Bonny's name has disappeared in legends of the Caribbean, as well as on the American mainland, her pirate counterpart, that other Irish Amazon, Grace O'Malley, has left her mark in her country throughout the landscape. Pocketed in the town of Westport on the west coast of Ireland there are Grace O'Malley Tearooms, occasionally Grace O'Malley Festivals and the omnipresent names of the O'Malleys of the past and present.

Clew Bay has perpetuated the legend, peculiar to many areas, of having at least 365 islands in its harbor. Whether or not there is an island for every day of the year, there is certainly a story for every island and around Clew Bay every story concerns Grace O'Malley. There are those who say that next to Queen Maeve, she was the greatest of all the Irish heroines; certainly she was the most colorful.' There is little about her in the national annals of Ireland, but in the state papers she does appear, known then by her Irish name Grainne Ni Mhaille and converted into Grace O'Malley because it was, as the old historians used to say, "the polite English form."

If Anne Bonny's father had been a respectable member of his community, Grace's father was more than respectable; indeed he was a chieftain of the Barony of Murrisk and, satisfied with his superb rank in life, married an O'Malley from the same county and family.

The beautiful, mythological genealogy of Ireland traces the O'Malley name back to time immemorial, a dusty landscape of the past where

Harbor seal, the best-known species. (Nineteenth-century print from the authors' collection) "They are as big as Calves, the Head of them like a Dog, therefore called by the *Dutch* the *Sea-hounds*. . . . Here are always Thousands, I might say possibly Millions of them, either sitting on the Bays, or going and coming in the Sea round the Island; which is covered with them (as they lie at the Top of the Water playing and sunning themselves) for a Mile or two from the Shore. When they come out of the Sea they bleat like Sheep for their Young; and tho' they pass through Hundreds of others' young ones, before they come to their own, yet they will not suffer any of them to suck." (From *A New Voyage Round the World* by William Dampier)

the O'Malleys had always been lords of Clew Bay. They were descended from the brother of Brian, the great progenitor of Connaught. Grace herself has left documents that tell us that her father's kingdom and then her own included all the islands of the surrounding ocean from Clare to Innisboffin, so that she was indeed not only a queen of the west, but the queen of a small kingdom of islands.

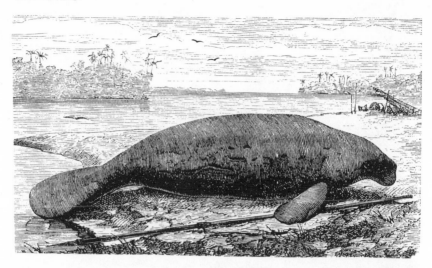

Manatee. (Nineteenth-century print from the authors' collection) "The Skin of the Manatee is of great use to Privateers, for they cut them into Straps, which they make fast on the Sides of their Canoas thro' which they put their oars in rowing, instead of Tholes or Pegs." (From *A New Voyage Round the World* by William Dampier)

In those days children were put out to foster and although we do not know where she was born, most think it was Belclare, one of the chief family castles. She was supposedly raised on her family's island, Clare, where the ever surrounding waters made her a seafaring genius, "the greatest captain in the western seas." She had to make a suitable marriage and appropriately enough she married Donall O'Flaherty of the great O'Flaherty clan of Connemara. It was they who controlled such islands as the Arans. Grace became as intimately familiar with these waters as with her own island territory, and her ship, they say, bore the sea horse of O'Malley and the lions of O'Flaherty.

As always there was trouble in Ireland and the English yoke grew tighter particularly in the west. Up until this period Grace O'Malley, in her own words, had been loyal to Queen Elizabeth, but now that loyalty was to be completely vanquished. Her son, her favorite son, Owen, was killed by Capt. John Bingham, brother of the infamous Sir Richard Bingham, as Owen was attempting to victual a ship of his mother's fleet on one of the family's islands. At the instigation of the British, Grace's family islands were raped of 4,000 cows, 500 brood mares, and 1,000

Coconut palm; nuts in various stages of growth. (Nineteenth-century print from the authors' collection) "The Coco-nut Trees grow by the Sea, on the Western-side in great Groves, three or four Miles in length, and a Mile or two broad. This Tree is in shape like the Cabbage-tree, and at a distance they are not to be known each from other, only the Coco-nut Tree is fuller of Branches; but the Cabbage-tree generally is much higher, tho' the Coco-nut Trees in some places are very high." (From *A New Voyage Round the World* by William Dampier)

sheep, a loss that was to leave the islanders in a state of abject poverty. Eighteen of Owen's followers were hanged and Grace, loyal no longer to the queen turned pirate.

Clare Island was her stronghold; her castle still stands there guarding the harbor and the beach below so that she could watch the galleys, always provisioned, always ready for pursuit of any enemy. Her name still lingers on all the islands of that shore because she needed, she felt, more than one stronghold, and had perhaps more than a dozen of them. She held utter command of Achill Sound and Clew Bay.

Her husband, who appears little enough in her story, died and she married again. Sir Henry Sydney has left a record that when he traveled to Galway in 1576 he was visited by a "most feminine sea captain called Grainne Ni Mhallie"; that she was married to a man called Richard in Iron, one of the great Burkes; and that she was now entitled to be called Lady Burke. She herself did not think of her deeds as piracy; she rather held what she called "maintenance by land and sea" of western Ireland. Knowing every twist of that indented coast, she could sail for protection into creeks and channels that were unfamiliar to her enemies. Her hus-

The first indication of the "greatest treasure of them all," the coral-covered wood and coins discovered by William Phips.

98

Clew Bay, the haunt of Grace O'Malley.

band warred mostly on the land; she warred mostly on the sea.

She was a great diplomat, too, and had no reluctance to try to keep up satisfactory relationships with Queen Elizabeth's emissaries. At the same time she did not hesitate to turn her back and attack whatever British ship might come within her reach.

Sometimes she found it worth her while to serve England, and it was inevitable that her name should be well known to Elizabeth at court, where Grace had sent emissaries informing Elizabeth that the Queen of the West would see the Queen of Hampton Court. She petitioned the Queen for land that she felt was hers after the death of her second husband. In return, said our wily Grace, she would prosecute the enemies of the British Queen with fire and sword. Her story is inscribed not only on the landscape of Ireland, but in the pages of her literature. She appears as misty as the landscape itself in James Joyce's great *Finnegans Wake* and sentimentally in many poems such as the following by Sir James Ferguson, extolling her exploits:

> Delighting on the broad-back'd deep,
>   To feel the quivering galley
> Strain up the opposing hill, and sweep
>   Down the withdrawing valley.
> Or, sped before a driving blast,
>   By following seas up-lifted,
> Catch, from the huge heaps heaving past,
>   And from the spray they drifted,
> And from the winds that tossed the crest
>   Of each wide-shouldering giant,
> The smack of freedom and the zest
>   Of rapturous life defiant.
>
> Sweet when crimson sunsets glow'd
>   As earth and sky grew grander,
> Adown the grass'd unechoing road,
>   Atlanticward to wander,
> Some kinsman's humbler heart to seek
>   Some sick bedside, it may be,
> Or, onward reach, with footsteps meek,
>   The low, grey, lonely abbey.

99

If it seems that the history of piracy did nothing but add color and excitement to the dusty pages of old tomes that we can now reread with great pleasure, the facts are more important. Indeed it was a pirate, William Dampier, more naturalist than pirate, more scientist than buccaneer, who opened the world's eyes to the wonders of different flora and fauna, yes, and of peoples around the world. Pirate he was, but travel writer par excellence far more. And it was his superb journal, which he carried 'around from place to place, its pages protected in a piece of bamboo, that guided even such great explorers as Capt. James Cook to unknown parts of the globe.

William Dampier was born in East Coker—immortalized by T. S. Eliot in *Four Quartets*—in the midseventeenth century. On the death of his parents he was apprenticed to a shipmaster and soon was off to the high seas making a voyage first to France and then to distant Newfoundland, a land far too cold for Dampier's more tropical tastes. It was the warmer climates that called him and he shipped aboard once again outward bound on an East Indiaman for Java. There he found the tropical world greatly to his liking and when upon his return an opportunity came to go out to manage an estate in Jamaica he settled there for a while as a planter. It was too quiet a life for him. He returned to England, married, returned to Jamaica to meet two remarkable buccaneers, Coxon and Sharpe, and turned buccaneer himself.

From there on, poor as he might have been as a buccaneer—Byron called him "the mildest mannered man that ever scuttled ship or cut a throat"—he collected information, if not gold, about the world. He kept his eyes open throughout all the Spanish Main, and the lure of the South Seas was for him less the lure of piracy going on there at the moment than of the infinite data that he scribbled down in his notebook. As a naturalist he observed carefully the Galápagos Islands, which were to be so important to Darwin. "I do believe," he said, "that there is no place in the world that is so plentifully stored with these animals." He was speaking of the giant tortoises. "The land turtle are here so numerous that five or six hundred men might subsist on them alone for several months without any other sort of provision." He made, too, a record of picking up an Indian named Will who had been left on the Juan Fernandez Island, an incident to be amalgamated into *Robinson Crusoe* by Daniel Defoe, who embroidered Alexander Selkirk's story. Selkirk had been picked up in these same islands by Dampier in a ship captained by Woodes Rogers before the latter decided to clean up New Providence.

He was an extraordinary island lover, and his observations, contained in a *New Voyage Round the World*, resulted from three such trips to a horde of islands, with the result that his book became one of the most influential of all times. He became respected by the scientists of his age and was almost idealized by a generation of poets, particularly Wordsworth and Coleridge, who found him a source of great inspiration. "Old Dampier," said Coleridge, "a rough sailor, but a man of exquisite mind"—the type of mind even more exquisitely exemplified by our next island lovers, Charles Darwin and Alfred Wallace.

100

101

Padre Island, the famous pirate island of Texas. (Courtesy Texas Highway Department)

6

Islands of Discovery

"THE VOYAGE of the *Beagle* has been by far the most important event of my life, and has determined my whole career; it had depended on so small a circumstance, my uncle offering to drive me thirty miles to Shrewsbury, which few uncles would have done, and on such a trifle as the shape of my nose. I have always felt that I owe to the voyage the first real training or education of my mind." So wrote Charles Darwin, describing what lay behind one of the greatest voyages of all time, his own voyage in the naval ship *Beagle* on a survey trip to the southern hemisphere that would revolutionize thought.

The immediate family had been against such a journey. It was his uncle's driving to Shrewsbury that meant that the final parental objection had been overcome. The shape of his nose (the ship's captain thought it proved lack of stamina) was offset by his obvious enthusiasm.

It is easy enough to understand what were some of the parental objections. "You care for nothing," his father said, "but shooting dogs and catching rats and you will be a disgrace to yourself and your family." Dr. Robert Darwin, the father of the genius of discovery, was a harsh taskmaster. He towered above his children and was addicted to a two-hour daily dissertation on the evils of the world and mostly those expressed in the character of his children. Young Charles naturally rebelled, but always quietly and always somewhat abjectly, and finally the total rebellion could come only in print. The young Charles Darwin was a creature of the early nineteenth century. He was born in 1809, when the world was making rapid shifts, when the agricultural economy shifted to urban, and a strange stage was being set, a stage wherein scientists would be the actors in the role of changing human thought.

There seemed little in Darwin's heritage or education that would fit him for the enormous contribution he was to make. True, it was a good family, a very good one. His grandfather had been *the* Dr. Erasmus Darwin, a man of many talents quite like Benjamin Franklin, and his other grandfather was the potter Josiah Wedgwood, a man of extraordinarily broad interests. Dr. Erasmus Darwin had written a philosophical poem on zoology, and the family's interests were quite strong in the world of natural history.

The Darwins were not unlike most of the good British squires of the period. Solitary walks, fishing, birds, and hunting were a part of daily life. Young boys did, of course, collect birds' nests, but they did consider the birds, and Josiah Wedgwood had even gone so far as to explain to young Charles the cruelty of dissecting a worm unless one first killed it in a bath of salt and water.

The boy read copiously, not always scientific work, but Shakespeare, Byron, and Scott and then the strange little compendium called **Wonders of the World**, one of the most unusual books of travel and exploration of the day. "I believe," Darwin said later, "that this book first gave me a wish to travel in remote countries which was ultimately fulfilled by the voyage of the *Beagle*."

But he was doing poorly in school, at least his father thought so, and so did his schoolmates, who nicknamed him "Gas" because of the amount of enjoyment he obtained from fooling around in the chemistry laboratory. Something had to be done and Darwin was sent to Edinburgh in October, 1825, to study medicine, following once again his father's footsteps. He did not make it. The sight of an operation made him ill and after two years another decision had to be made. His father made it for him: He was going to Cambridge to study for the ministry. He entered Christ College in the autumn of 1827 and in the spring of 1831 obtained his B.A. He had disliked classwork enormously, but he had come in touch with many naturalists and was already an amateur collector of note.

He had been collecting insects since the age of ten when the family would vacation on the seashore in Wales. Beetles in particular fascinated him. When he undressed they would fall out of his clothes. When he would run across an unusual specimen he was ecstatic. One day in the bark of an old tree he was able to find two extremely rare beetles. He held them one in each hand, then he grew utterly bewildered. Having

103

104

Darwin's study.

captured the two, he could see on the tree another even rarer. He solved the problem by putting the third beetle in his mouth. Soon this extraordinary beetle collector came to the attention of older scientific men at college: Professor J. H. Henslow, the naturalist, and Leonard Jenyns, the zoologist.

He learned taxidermy, trained by the same man who had taught Charles Waterton, the great explorer. His beetle collection even brought him into print. "No poet," he said, "ever felt more delighted at seeing his first poem published than I did at seeing in Stevens' *Illustrations of British Insects,* the magic words captured by C. Darwin, Esq." So a beetle perhaps more than anything else led him to the opportunity to make the voyage that was to make history.

He was ready when the opportunity presented itself, a great opportunity. Captain R. Fitzroy of the Royal Navy, who was scientific enough to judge a man by the shape of his nose, was at least enthusiastic enough to know that on the voyage that he was to undertake he should have with him a naturalist to study the flora and fauna. He called upon

Orangutan attacked by Dyaks.

his friend Professor Henslow, who told him about Darwin, who had just finished at Cambridge. After the family objections were settled, Darwin wrote to the captain, "What a glorious day the fourth of November will be to me. My second life will then commence, and it shall be as a birthday for the rest of my life." It was, but the birthday started poorly enough on the *Beagle*. Known in those days as a coffin, the ship had a crew of thirty-five. A bark of two hundred and thirty-five tons, it was a type that had a bad reputation for sinking in a gale. She was not to sink with Darwin aboard, fortunately, but she was tossed and twisted like a proverbial sea monster, and Darwin in those tedious years was to be constantly ill. He would work for only about an hour and then he would "take the horizontal," stretch himself out and hope that he would die. He never got his sea legs and he suffered throughout the voyage, but at each island at which the ship stopped he would pull himself together, race down the gangplank and begin his collecting. He chipped rocks, collected lichens, spiders, moths, and observed every bird. All of the long voyage was important to him; every island he stopped at fascinated him.

But two groups of islands in specific made their way into scientific literature; one set was the Galápagos, and the other the coral islands of the Pacific.

On Chatham Island he found almost immediately the giant tortoises (the Galápagos had been named for them by the Spanish), and he was fascinated in following the strange trails which they made crawling for water in the interior. He tested their speed. They went at least, he found, four miles a day. He began to wonder deeply. "The archipelago," he wrote, "is a little world within itself, or rather a satellite attached to America whence it has derived a few stray colonists and has received the general character of its indigenous production. Considering the small size of these islands we feel the more astonished at the number of their aboriginal beings and at their confined range. Seeing every height crowned with its crater and the boundaries of most of the lava streams still distinct, we are led to believe within the period geologically recent the unbroken ocean was here spread out, hence in space and time we seem to be brought somewhere near to that great fact—that mystery of mysteries—the first appearance of new beings on this earth." In effect his great book, THE ORIGIN OF SPECIES, had begun.

106

Six months later the *Beagle* was in Keeling Island in the Pacific. Keeling was an atoll and the coral reefs facinated him. Here he found the giant land crab *Birgus* that lived on coconuts, first stripping the husk, then inserting its fifth claw with pincers into the nut. Says Darwin: "To show the wonderful strength of the front pair of pincers, I may mention that Captain Moresby confined one in a strong tin box, which had held biscuits, the lid being secured with wire; but the crab turned down the edges and escaped. In turning down the edges it actually punched many small holes quite through the tin!"

He began to brood about the structure of the reef. There were three types, he decided: atolls, barrier, and fringing. And he immediately began to wonder how such coral islands were deposited:

We must feel astonished at the vastness of the areas, which have suffered changes in level either downwards or upwards, within a period not geologically remote. It would appear, also, that the elevatory and subsiding movements follow nearly the same laws. Throughout the spaces interspersed with atolls, where not a single peak of high land has been left above the level of the sea, the sinking must have been immense in amount. The sinking, moreover, whether continuous, or recurrent with intervals sufficiently long for the corals again to bring up their living edifices to the surface must necessarily have been extremely slow. This conclusion is probably the most important one which can be deduced from the study of coral formations; and it is one which it is difficult to imagine . . . Nor can I quite pass

Mt. Cella and Agostini Fjord near Tierra del Fuego, Chile. (Courtesy Grace Line)

over the probability of the former existence of large archipelagoes of lofty islands, where now only rings of coral-rock scarcely break the open expanse of the sea, throwing some light on the distribution of the inhabitants of the other high islands, now left standing so immensely remote from each other in the midst of the great oceans. The reef-constructing corals have indeed reared and preserved wonderful memorials of the subterranean oscillations of level; we see in each barrier-reef a proof that the land has there subsided, and in each atoll a monument over an island now lost. We may thus, like unto a geologist who had lived his ten thousand years and kept a record of the passing changes, gain some insight into the great system by which the surface of this globe has been broken up, and land and water interchanged.

The *Beagle* returned to England in 1836. It had been a long birthday party, nearly five years long. From it he was to make his own gift, a gift to the ages—a book that was to be called the most important of the century, *The Origin of Species*. In 1837 he began to write down his first remarkable contribution about the mutability of species. Seven years later he was writing to Joseph Hooker, who had just returned from an Antarctic voyages on the *Erebus*. "I am almost convinced . . . that species are not (it is like confessing a murder) immutable."

He continued to work on his book and his theory. He was shocked in 1855 to discover a paper by another island lover, Alfred Russel Wallace. Could it be that Darwin, who had been working twenty years on one great idea, was to discover that somebody else had also reached the same conclusion—and curiously enough through voyages to islands and a careful study of flora and fauna peculiar to islands?

Alfred Russel Wallace was born in England on January 8, 1823. He worked first as a land surveyor and architect and only indirectly became interested in botany. He soon made his own herbarium. While a schoolmaster he met the naturalist H. W. Bates and started a beetle collection. His career as one of the greatest naturalists of all times and a co-developer of the idea of the origin of species was under way. First he went to the Amazon collecting madly, but most of his great collection was lost enroute. In 1852 he made the trip to the Malay Archipelago that was to go down in history.

In his preface to the book entitled *The Malay Archipelago*, one of the great natural-history classics of all time and one that is still enchanting reading on a winter day, he says:

108

Unique in the kangaroo family, the ring-tailed rock wallaby lives in the rocks and uses its tail as a sort of rudder to guide it over difficult terrain. (Courtesy Australian Tourist Commission)

When I reached England in the spring of 1862 I found myself surrounded by a room full of packing cases containing the collections that I had from time to time sent home for my private use. These comprised nearly three thousand birdskins of about a thousand species and at least twenty thousand beetles and butterflies of about seven thousand species; besides some quadrupeds and land shells.

His entire collection on his archipelago trip, which lasted for eight years, contained 125,000 specimens. He had obtained them over a period of eight years by going 14,000 miles through the archipelago on foot and by native canoe.

Like Darwin he had what has been described as an alluvial mind; constant observation watered his insight until it grew into a river that was a source of knowledge. Unlike Darwin, who took twenty years to finally evolve his theory of the origin of species, Wallace had a much shorter and fiercer creative experience. In 1855 while staying in Borneo he wrote an essay which contained the words: "Every species has come into existence coincident both in time and space with a pre-existing closely allied species."

The echidna, or spiny anteater, of the Southwest Pacific is sluggish and seemingly stupid. The female develops a pouch during the breeding season in which one or two eggs are placed. (Courtesy Australian Tourist Commission)

A *haus tambaran*—secret meeting place for men only—which was erected at the 1965 Mt. Hagan show, New Guinea.

110

His own observations began to haunt him even more persistently. He could never get them out of his mind from that point on. He said, "How changes of species could have been brought about was rarely out of my mind."

His trip into Borneo was to pay great dividends, but at the same time to almost destroy his health. He was to move cautiously through all the islands, first the Indo-Malay, then the Timor group, then the Celebes, the Moluccas and finally the Papuan group. The Malay Archipelago extends for more than 4,000 miles in length and is about 1,300 miles from north to south. Wallace kept comparing the size of the islands with his own Great Britain. All of the British Isles, for example, could be set down on the one island of Borneo.

Borneo in particular fascinated Wallace. It was, of course, the home of that strange animal the orangutan. Smaller animals and insects fascinated him equally. He doubled his insect collection in Borneo and was delighted to collect Borneo's large flying tree frog as well as many rare butterflies. But it was the mias, as the natives called it, or the orangutan, that fascinated him most. He was the first European to study it in detail. He was fortunate in locating a young mias whose mother had been killed.

Australia has some of the most unusual
island fauna in the world. The koala bear
is an arboreal marsupial. It feeds only
upon the buds and leaves of the eucalyptus
tree. (Courtesy Australian Tourist Commission)

111

Great turtle of the Galápagos Islands. These islands received their name from the
Spanish word *galápago*, which means "tortoise." (Nineteenth-century print from the
authors' collection)

He wondered what he should feed it; the Malays, the Chinese, and the Dyaks he knew had no milk and instead he had to suckle it with rice water in a bottle having a quill in the cork.

When handled or nursed, it was very quiet and contented, but when laid down by itself would invariably cry; and for the first few nights was very restless and noisy. I fitted up a little box for a cradle, with a soft mat for it to lie upon, which was changed and washed every day; and I soon found it necessary to wash the little Mias as well. It enjoyed the wiping and rubbing dry amazingly, and when I brushed its hair seemed to be perfectly happy, lying quite still with its arms and legs stretched out while I thoroughly brushed the long hair of its back and arms. For the first few days it clung desperately with all four hands to whatever it could lay hold of, and I had to be careful to keep my beard out of its way, as its fingers clutched hold of hair more tenaciously than anything else, and it was impossible to free myself without assistance. . . . Finding it so fond of hair, I endeavoured to make an artificial mother, by wrapping up a piece of buffalo-skin into a bundle, and suspending it about a foot from the floor. At first this seemed to suit it admirably, as it could sprawl its legs about and always find some hair; which it grasped with the greatest tenacity. I was now in hopes that I had made the little orphan quite happy; and so it 112 seemed for some time, till it began to remember its lost parent, and try to suck. It would pull itself up close to the skin, and try about everywhere for a likely place; but, as it only succeeded in getting mouthfuls of hair and wool, it would be greatly disgusted, and scream violently, and after two or three attempts, let go altogether. One day it got some wool into its throat, and I thought it would have choked, but after much gasping, it recovered, and I was obliged to take the imitation mother to pieces again, and give up this last attempt to exercise the little creature. . . .

When I had had it about a month, it began to exhibit some signs of learning to run alone. When laid upon the floor it would push itself along by its legs, or roll itself over, and thus make an unwieldy progression. When lying in the box it would lift itself up to the edge into almost an erect position, and once or twice succeeded in tumbling out. When left dirty, or hungry, or otherwise neglected, it would scream violently till attended to, varied by a kind of coughing or pumping noise, very similar to that which is made by the adult animal. If no one was in the house, or its cries were not attended to, it would be quiet after a little while, but the moment it heard a footstep would begin again harder than ever.

Wallace tended his pet carefully, but within several months it lost its appetite completely and died.

If the large orangutan interested him, so did the smaller flora and fauna—the durian, for example, a local fruit which had been known to fall on top of a man and kill him. He decided that if the orange was the queen of fruits, then this tropical delicacy was the king. Like so many of the tropical fruits, alas, it had an immediate smell that was unfavorable. The old travelers in 1599 who had run across it described it simply as a rotten onion. But Wallace would try anything, from this fruit to walking the precarious bamboo bridges. He was fascinated by bamboo as are all travelers to tropical islands, and noted the uses to which the Dyaks put it. He went out each night with some of the Dyaks to watch them climb the branches of the tappan tree, whose cylindrical trunk juts a hundred feet into the air all smooth and branchless. Great bamboo stalks were driven into the ground, and the Dyaks made a bamboo ladder as they climbed to reach the tops of these trees. Bamboo had all sorts of other uses—baskets, hen coops, birdcages, and fish traps. Water was carried to the homes by making little aqueducts of the large bamboo split in half and then supported by crossed sticks of various heights. Bamboo, too, served for cooking utensils. Fish was preserved in bamboo containers; rice was boiled in them; knives had bamboo sheaths.

Everything from pitcher plants to ferns attracted him, and he tried to catch every moth he saw.

He made long and careful notes about the aborigines, the Dyaks. A true gentleman of his time and an almost anthropologist, he was concerned with the morals of each aboriginal group he would meet and the Dyaks impressed him with their character:

113

> The moral character of the Dyaks is undoubtedly high—a statement which will seem strange to those who have heard of them only as head-hunters and pirates. The Hill Dyaks of whom I am speaking, however, have never been pirates, since they never go near the sea; and head-hunting is a custom originating in the petty wars of village with village, and tribe with tribe, which no more implies a bad moral character than did the custom of the slave-trade a hundred years ago imply want of general morality in all who participated in it.

But even in the heart of the jungle he thought well. He was reading Malthus's book *An Essay on the Principle of Population*, applying it to the native tribes, and it was during an evening of such observation that there "suddenly flashed upon me the idea of the survival of the fittest."

In a fit of fever he wrote up his observation and sent it to Charles Darwin. That observation was to make scientific history, but Wallace's collecting days were by no means over. While Darwin sat at home brooding about the fact that Wallace in the deep jungle had come up with the same idea that he had laboriously traced out since his days on the *Beagle* and now thrashed out with himself in Down House, Wallace continued his adventures.

ISLANDS

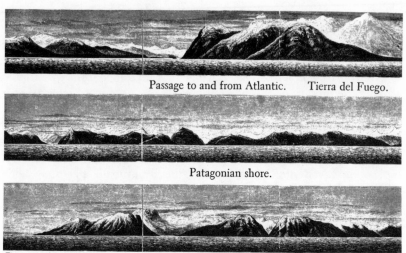

Passage to and from Atlantic.   Tierra del Fuego.

Patagonian shore.

Entrance to Pacific.   Exit from Magellan Strait.

Charles Darwin made one of the first scientific expeditions through the Strait of Magellan.

Racked with fever, often hungry, he continued his journeys. Sometimes it would be just chance that he would go to one place instead of another. Once when it was impossible to go to Singapore, he turned instead to Bali and Lombok, thereby making some of the most important discoveries of his whole expedition to the East. There he made his way through thorny jungles. Even the bamboos were thorny. He lived in one room eating, sleeping, working, and dissecting. The room had no chairs or tables; armies of ants covered his working space; wild dogs came in at random. He evolved the method of putting an old four-legged bench in coconut shells filled with water. In this way he kept some of his material away from the ants. He moved on from island to island, always meeting with unusual difficulties:

One morning, as we were sitting at breakfast, Mr. Carter's servant informed us that there was an "Amok" in the village—in other words, that a man was "running a muck." Orders were immediately given to shut and fasten the gates of our enclosure; but hearing nothing for some time, we went out, and found there had been a false alarm, owing to a slave having run away, declaring he would "amok," because his master wanted to sell him. . . . In their wars a whole regiment of these people will sometimes agree to "amok," and then rush on with such energetic desperation as to be very formidable to men not so excited as themselves. Among the ancients these would have been looked upon as heroes or demigods who sacrificed themselves for their country. Here it is simply said—they make "amok."

Macassar is the most celebrated place in the East for "running a muck." There are said to be one or two a month on the average and five, ten, or twenty persons are sometimes killed or wounded at one of them. It is the national and therefore the honourable mode of committing suicide among the natives of Celebes, and is the fashionable way of escaping from their difficulties.

From March to July of 1858 he made his long-anticipated trip to New Guinea. He who was a great collector was also a great planner, and knowing that he would have to build a house, he took with him eighty waterproof mats of pandanus leaves which would roof his structure. The voyage was by no means easy—monsoons appeared regularly, then calms and always contrary currents. Natives in canoes would come to the schooner carrying coconuts, pumpkins and palm-leaf mats, but they annoyed Wallace because they had been dealing with the whalers on China's ships who were used to paying them "extravagantly." If a yard of calico would not do, Wallace would have little from them. During the entire voyage he contained within himself a muted excitement. This was the country of the cassowary and the tree kangaroo, but more impor- 115 tantly the dark forests of this island produced, as he said himself, "the most extraordinary and the most beautiful of the feathered inhabitants of the earth—the varied species of birds of paradise."

When he landed he overcame all sorts of obstacles and managed to build a satisfactory house, only then to get a severe infection in his foot, the kind that had bothered him and would bother him throughout all his voyages. Leeched and lanced and doctored, he was, as he said, almost driven to despair. He would watch birds fly by that he was likely never to see again, butterflies that the world had never known before. But occasionally some of the natives would bring him artwork—what we now know was some of the most extraordinary artwork of the world, the carvings of New Guinea.

He commented that the bird of paradise often eluded him. It took him five voyages to get five species of the fourteen birds of paradise known to exist in the New Guinea district. Nonetheless, he was the only Englishman to have seen these birds in their native forest. These birds had haunted the voyagers to the islands since the very earliest travelers in search of cloves and nutmegs. Sometimes the dried skins would be given to a Malay trader. They were called God's Birds, and the Portuguese had named them Birds of the Sun. The early naturalists of the sixteenth century, none of whom had seen the birds alive, said that they

116

The interest is mutual between a very young girl and a New Guinea tribesman. The tribal cultures of Australia, and indeed the mores of all the Pacific islands, have attracted the world's great anthropologists.

lived in the air, always turning toward the sun, and did not light on the earth until they died, for they had neither feet nor wings. Naturally the old dried specimens rarely had wings, and the natives never preserved the feet.

Wallace was the first naturalist to hear that *wauk, wauk* which is the cry of the bird and he was never to forget it. He was so taken with them that he took two back with him to England, feeding them carefully on the trip back, stopping in Bombay for bananas and cockroaches in Malta and protecting them from the cold winds of the Mediterranean in March. They lived in the Zoological Park for only a couple of years. Wallace considered his discovery of the paradise birds one of his most exciting contributions to natural history. It was, however, his island studies of the evolution of species that proved his genius. He and Darwin were the fathers of the thousands of scientists who have found in the isolation of island life, flora and fauna, a true laboratory of scientific discovery.

117

7

The Call of the South Seas

"CALL ME ISHMAEL," and send me to the South Seas. The call of the
South Seas has been persistent since the discovery of that "almost Eden."
Throughout the old documents runs a constant refrain that there, in the
land of breadfruit trees and easy passion, the most Calvinistic New Englanders, the most determined Scots, the most phlegmatic Englishmen
became something else, something different and abandoned; life took on
a fresh new beauty; there was a world of love and leisure, almost to be
had for the asking.

Certainly the world's writers did much to encourage such superb
fantasies and perhaps some of the best, or at least the most attractive,
pictures of island life come from the writers. Robert Louis Stevenson,
for example, pining for health and the comfort of the sun, sailed through
many lagoons and navigated many reefs to try to find some place in the
Pacific to call home—a place far from the Edinburgh where he had been
born.

Having discovered the pleasures of the islands, Stevenson and his
wife once chartered an iron screw cargo boat, a topsail schooner of some
six hundred tons gross, named the *Janet Nichol*. Surely there couldn't
have been a life more distant, more distinct from the gray smog of Edinburgh. Here on the deck the Stevensons would sit and sleep, sometimes
in a hammock or frequently on a mat. The Stevensons came across
strange little islands, encountering strange customs with the ease of an-

thropologists on a spree. Sometimes the very sailing ventures were perilous, and Stevenson wrote many letters home recording the voyage:

Honolulu, Hawaiian Islands, February, 1889.

My dear Bob—My extremely foolhardy venture is practically over. How foolhardy it was I don't think I realized. We had a very small schooner, and, like most yachts, over-rigged and over-sparred, and like many American yachts on a very dangerous sail plan. The waters we sailed in are, of course, entirely unlighted, and very badly charted; in the Dangerous Archipelago, through which we were fools enough to go, we were perfectly in ignorance of where we were for a whole night and a half the next day, and this in the midst of invisible islands and rapid and variable currents; and we were lucky when we found our whereabouts at last. We have twice had all we wanted in the way of squalls; once as I came on deck, I found the green sea over the cockpit coamings and running down the companion like a brook to meet me; at that same moment the foresail sheet jammed and the captain had no knife; this was the only occasion on the cruise that ever I set a hand to a rope, but I worked like a Trojan, judging the possibility of hemorrhage better than the certainty of drowning. . . . What was an amazement, and at the same time a powerful stroke of luck, both our masts were rotten, and we found it out—I was going to say in time, but it was stranger and luckier than that. The head of the main-mast hung over so that hands were afraid to go to the helm; and less than three weeks before—I am not sure it was more than a fortnight—we had been nearly twelve hours beating off the lee shore of Eimeo (or Mooréa, next island to Tahiti) in half a gale of wind with a violent head sea: she would neither tack nor wear once, and had to be boxed off with the main-sail—you can imagine what an ungodly show of kites we carried—and yet the mast stood. The very day after that, in the southern bight of Tahiti, we had a near squeak, the wind suddenly coming calm; the reefs were close in with, my eye! what a surf! The pilot thought we were gone, and the captain had a boat cleared, when a lucky squall came to our rescue. My wife, hearing the order given about the boats, remarked to my mother, "Isn't that nice? We shall soon be ashore!" Thus does the female mind unconsciously skirt along the verge of eternity. Our voyage up here was most disastrous—calms, squalls, head sea, waterspouts of rain, hurricane weather all about, and we in the midst of the hurricane season, when even the hopeful builder and owner of the yacht had pronounced these seas unfit for her. . . .

From my point of view, up to now the cruise has been a wonderful success. I never knew the world was so amusing. On the last voyage we had grown so used to sea-life that no one wearied, though it lasted a full month, except Fanny, who is always ill. All the time our visits to the islands have been more like dreams than realities: the people, the life, the beachcombers, the old stories and songs I have picked up, so interesting; the climate, the scenery, and (in some places) the women, so beautiful. The women are

119

View of Tahiti.

120

handsomest in Tahiti, the men in the Marquesas; both as fine types as can be imagined. Lloyd reminds me, I have not told you one characteristic incident of the cruise from a semi-naval point of view. One night we were going ashore in Anaho Bay; the most awful noise on deck; the breakers ·distinctly audible in the cabin; and there I had to sit below, entertaining in my best style a native chieftain, much the worse for rum! You can imagine the evening's pleasure.

One stirring day was that in which we sighted Hawaii. It blew fair, but very strong; we carried jib, foresail, and mainsail, all single-reefed, and she carried her lee rail under water and flew. The swell, the heaviest I have ever been out in—I tried in vain to estimate the height, at least fifteen feet—came tearing after us about a point and a half off the wind. We had the best hand—old Louis—at the wheel; and, really, he did nobly, and had noble luck, for it never caught us once. At times it seemed we must have it; Louis would look over his shoulder with the queerest look and dive down his neck into his shoulders; and then it missed us somehow, and only sprays came over our quarter, turning the little outside lane of deck into a mill race as deep as to the cockpit coamings. I never remember anything more delightful and exciting. Pretty soon after we were lying absolutely becalmed under the lee of Hawaii, of which we had been warned: and the captain never confessed he had done it on purpose, but when accused, he smiled. Really, I suppose he did quite right, for we stood

One of a series of twelve crystal plates engraved to depict collectors' shells gathered from the seven seas. The shell depicted here is the slit shell. Designed by Don Wier. (Courtesy Steuben Glass)

121

committed to a dangerous race, and to bring her to the wind would have been rather a heart-sickening manoeuvre.

R. L. S.

And what a charming record Fanny Stevenson kept of that extraordinary trip, not for her the musing of her husband, but just the practicality of a woman making her way as though she headed some great native procession on distant coral atolls. Fanny, for example, having discovered that there were few flowers in the atolls and that the natives enjoyed them so much, simply made her own. She sent to Sydney for boxes of old-fashioned flowers which she wired into wreaths to be bestowed upon the natives. There was Fanny waiting desperately for the weather to turn warm enough so that she could go barefooted on the slippery decks of the ship, only to discover that the chill even demanded flannel bodices; Fanny looking for onions and discovering that the only place that she could obtain them was where they were planted on the graves; Fanny breaking up one of the native arguments with her peace stick—one of those sticks that could be waved in the center of any group, and instant peace, a curious pleasure in this uncivilized world, would descend.

ISLANDS

The islands fascinated Fanny, and their names became as familiar as the names of towns and villages in which she had grown up in California. Even in those days, or perhaps more in those days than in any other time, the beachcombers were making their mark and their legends in the South Seas. King Jennings, for example, an American who had married a Samoan, lived now on the island of Queros in a little kingdom of his own with a flag portraying a dove in a field. "We asked with some curiosity what the dove indicated," she wrote; "they told us that a night bird came and cried about the settlement for months; this was supposed to bode sickness, so to propitiate the ill-omened bird, it was added to the flag."

She sympathized with the islanders who came from other islands to work in the fields—the "labor boys" as they were called—a practice that still occurs not only in the South Seas, but in the Caribbean, where an islander must leave his island for a better livelihood. The labor boys were often depressed. They always had the same symptom—"just plain homesickness for a cannibal island." Fanny remarked that it did not matter that mother and father and all the rest had been killed and eaten long ago; more than one labor boy would die of plain old homesickness.

She recorded the miserable account of the sick native whose family ignored him completely and simply started to make a coffin. "But the man is not dead," she argued. "Oh, yes," was the reply, "he's dead enough; it's the third time he has done this, so we are going to bury him." Despite the fact that the man was in convulsions and still alive, they would accept only the fact that his spirit had long since departed from his body. When he became too cantankerous they simply took a large beam and quieted him once and for all.

If there were dreams in the islands, there were the realities too: the illness, the half-castes with elephantiasis, the leprosy, the general ennui. Often the Stevensons would ask someone who had gone to the islands for a new life what he thought of his choice. "It is no damn good at all," was the reply.

As the Stevensons sailed, Louis dictated to his stepson Lloyd. The *Janet Nichol* became like some *Flying Dutchman* inhabited by the ghosts of an Edinburgh past, making their way to the South Seas. On January 20, 1890, Stevenson wrote to the doctor who had taken such excellent care of him when he had been an invalid at Bournemouth, that he had been in the South Seas twenty months and he was, he felt, a person whom the doctor would scarcely know. "I think nothing of long walks and rides," he wrote; "I was four hours and a half gone the other day, partly riding, partly climbing up a step ravine."

Delighted with his own physical well-being, he decided to settle in Samoa and bought three hundred and fourteen and a half acres in the bush behind Appia. So began one of the greatest island legends of all time—Stevenson's own, the tale of Tusitala the storyteller in Samoa, part planter, part settler, an almost king, a politician, a writer, an almost god.

From November, 1890, until he died Stevenson was all these things, writing, talking, living his own fantasies. Life was not always ideal. Occasionally they even went hungry, Fanny and Louis sharing one avocado pear or Louis dining on hard bread and onions. It was too soon for guests. What would they do to feed them, he once wrote, serve up a labor boy fricasseed? He wrote well, scratching out endless data about the islands themselves, occasionally getting books from home. Who was this new man Kipling, for example? "By far the most promising young man who has appeared since, ahem, I appeared."

But this man who loved islands, the man who created the fantasy of *Treasure Island*, was, as his health grew poorer, drawn to another fantasy—the distant, misty fantasy of the Scotland that he had left behind with its fog-bound islands that he had known. The South Seas awakened a passion of memory as he started *The Weir of Hermiston*, the greatest of all his books.

123

His letters from Samoa took on a new bitterness. "I shall never take that walk by the Fisher's Tryst in Glencourse. I shall never see old Reekie. I shall never set my foot again upon the heather. Here I am until I die, and here will I be buried. The word is out and the doom written."

But life, certainly, had grown much better. There was now at Vailima an almost lavish hospitality; the great redwood hall was filled with ceremonial dinners. Stories began to appear in the press, far away in England or America, that the Stevensons lived a feudal life. He was a chief among chiefs in the islands, an almost king in a kind of Arabian Nights entertainment. Robert Louis Stevenson in the Pacific Islands, Robert Louis Stevenson in the South Seas—the stories were legion. The pictures appeared in the papers too; Stevenson "negligently attired," they called it in those days, in a sleeveless shirt, trousers up to the knee, bare feet. One September there were new pictures of him in the paper and the story about "the road of the loving heart." The Samoans had connected his house with a public road, farther down the hill, as a free gift. It was one of the greatest of all compliments and on its side was placed a board with an inscription that concluded: "It shall never be muddy, it shall endure forever, this road which we have made."

Stevenson did not walk it long: He died in December, 1894.

If such writers as James Norman Hall and Charles Nordhoff as well

124

A Trukese fisherwoman, interrupted by a photographer, smiles while ready to throw the net. (Courtesy Trust Territory of the Pacific Islands)

Plantation worker near Rabaul clearing palms affected by the rhinoceros beetle. (Photograph by Charles Barrett)

A Trukese assists in preparing preserved breadfruit for consumption during off-season. (Courtesy Trust Territory of the Pacific Islands)

125

as Somerset Maugham were to be the literary sons of Stevenson, the grandfather of all South Sea island writers was, of course Herman Melville. With those ringing words "Call me Ishmael," he started on a fresh new course in American literature, and in the pages of *Typee* and *Omoo* about his travels in the South Seas he opened up a rich new world.

The world had been opened up before, of course, by those sailing captains of such places as Salem and Boston and Maine who knew the far-off islands of the Indies and those of the far Pacific as well as they knew Boston Common or the Salem Athenaeum. Indeed the Peabody Museum in Salem to this day is almost caught in time—the time of the great sailing days to the Indies and the Caribbean. There is nothing more fascinating than to go into the back rooms of any museum, the Peabody perhaps more than any, and see secreted in the drawers material that is yet to be displayed: strange birds—including a folded albatross that came from the southern hemisphere, the gift of an old sailing captain—strange curios, china cups, and precious textiles. Once nearly every home in the seaports of both our coasts had these mementos of faraway sailing journeys when the call of the South Pacific was the call of trade and of that great leviathan of the sea, the whale.

So Melville, the grandfather of all the romanticism of the South Seas, said to call him Ishmael. He sent Ishmael on that great journey after the white whale, but the books before *Typee*, for example, had already given Melville the reputation as "the man who had lived among cannibals."

We now know that Melville spent shockingly little time in the South Seas, that the extraordinary amount of information that he jammed into his books had many aspects that were anthropologically, geologically and geographically impossible. Robert Louis Stevenson was to complain with enormous petulance that poor Melville, master though he was, simply did not know how to spell the Polynesian words. It was not a fact that would have bothered Melville too much. He was primarily an explorer of the unconscious. These worlds of islands were but a device to carry his characters through their lonely pilgrimage and exile of travel. His savages, "noble savages," were frequently borrowed, but many of his sources were impeccable, from Starbuck's whaling manuscript to earlier diaries and reports that he had discovered. So he gives us the Marquesas, those strange islands, creating an excitement that has not died down in any of us with those extraordinary lines: "Hurray my lads, it's a settled thing. Next week we shape our course to the Marquesas."

126

But over the years a controversy has raged, particularly about *Typee* and *Omoo* and *Mardi*. What was valid in those books? Had indeed Melville ever seen the Marquesas? He had indeed, although he sometimes borrowed his information.

Omoo, the island rover, is perhaps the most autobiographical of Melville's books and in it he gives some of the feeling of that time in the South Seas, projecting extraordinary pictures of early-nineteenth-century Tahiti, Honolulu and of beachcombing life, as well as the pathetic and degenerating state of the natives. Their prospects are hopeless, he said. Like other uncivilized beings brought into contact with Europeans they must remain stationary here until utterly extinct. And in one final painful roundup:

About the year (1767), Captain Cook estimated the population of Tahiti at about two hundred thousand. By a regular census, taken some four or five years ago, it was found to be only nine thousand. This amazing decrease not only shows the malignancy of the evils necessary to produce it, but from the fact the inference unavoidably follows that all the wars, child murders, and other depopulating causes, alleged to have existed in former times, were nothing in comparison to them.

These evils, of course, are solely of foreign origin. To say nothing of the effects of drunkenness, the occasional inroads of the small-pox, and other things which might be mentioned, it is sufficient to allude to a virulent disease which now taints the blood of at least two-thirds of the common people of the island; and, in some form or other, is transmitted from father to son. . . .

Distracted with their sufferings, they brought forth their sick before the missionaries, when they were preaching, and cried out, "Lies, lies! you tell us of salvation; and, behold, we are dying. We want no other salvation than to live in this world. Where are there any saved through your speech? We are all dying with your cursed diseases. When will you give over?"

He observed life at sea too, the terrible cruelty of the corporal punishment of our navy which he opposed in *White Jacket*. Then he wrote. He wrote in such a way that Hawthorne said you sometimes had to swim for your life just to read it. He sang of the sea and of the islands and of the people, of man's loneliness and inhumanity, bewilderment and despair, of man, the rover, who would never truly touch home.

So this, then, was one who could inspire anyone. Certainly James Norman Hall admitted Melville led him to the South Seas.

James Michener, who gave us his own *Tales of the South Pacific*, likes to say that it fascinated him that the greatest American novel should be not about America itself, but Herman Melville's white whale in the South Seas.

127

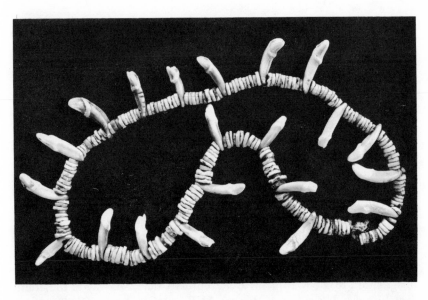

Canine teeth that served as coins; from the Solomon Islands. (Courtesy Chase Manhattan Bank)

128 Underwater close-up of the coral Great Barrier Reef in Australia. (Courtesy Australian National Travel Association)

Melville, then Jack London and, perhaps most important of all, James Norman Hall were contributors to the legend, lore, and literature of the South Seas. None had more affection for the islands of the Pacific than Hall. "Ever since boyhood," he wrote, "the mere name island has had a peculiar fascination for me." Islands were rare in his Midwest. "Islands were far to seek out on the prairies of Iowa and yet they could be found of a sort. A mud bank in the sluggish mid-stream of a prairie slough was enough; and if at the season of the spring rains I found one larger, with a tree or two, the roots undermined by the current, leaning across it, I had nothing better than to halt there and moor my flatbottomed skiff to the roots of one of the trees. Try as I would, though, I could not imagine the sea—any sea—the fact that the earth is three-quarters water was not a fact to me."

When he was about ten or eleven Hall discovered a book that was to change his life: *Typee* by Melville. It was the first time that he emotionally believed in the sea, and, as he said, he followed Melville in the imagination to the Marquesas and eventually made his way to the South Seas himself. There James Norman Hall and Charles Nordhoff were to start a fruitful collaboration and a strange literary partnership. For

"The Bathers" by Paul Gauguin. (Courtesy National Gallery of Art)

twenty-five years they lived and worked together on Tahiti, writing *Mutiny on the Bounty, Men Against the Sea, Pitcairn's Island, The Dark River,* and *The Hurricane.* Hall used to like to tell the story of arriving once in his Aunt Harriet's house after he had lived for a considerable length of time in the South Seas, only to be greeted with, "Now, dear, we must have a long talk. I want you to tell me why you live on that wretched little island. You must tell me what keeps you there."

So why did Hall live in Tahiti? The important thing, Hall felt, was for an individual to discover for himself the environment best suited to him and then to stick to it as long as he could. American life, he felt, was alien to him, moving too rapidly with little time for the pursuits he most loved. In an island world, he felt, all could be comprehended; social and political activity was right at hand; the complexity of life for once became possible to understand or evaluate.

And finally, of course, there is one of Britain's major contributors to the South Sea lore. Who will ever forget Joan Crawford swaying in the doorway behind that beaded curtain, the glorious evolution of a character, Sadie Thompson, in Somerset Maugham's *Rain?* Throughout his trips in the South Seas, Maugham kept a diary. Good novelists, he always pointed out, had to work from life. In that diary on one of his journeys from Honolulu to Pago Pago he had made notes that eventually evolved into *Rain*: the missionary with the long loosely jointed limbs and the hollowed cheeks, the plump pretty prostitute, the port of Pago Pago in south Samoa, the rain and the gray clouds and the boardinghouse with its verandas and with the rain rattling down on the corrugated roof.

Maugham always considered the South Pacific the most interesting of the world's oceans. He, like James Michener, was to be fascinated by the volcanic activity in all the islands, particularly the Kilauea volcano on the big island of Hawaii. Stupendous and alarming he called it, like some huge formless creature born of primeval slime.

He tracked down all legends of Robert Louis Stevenson, Pierre Loti, Rupert Brooke, the latter having been named "Pupure, the blond one" by the Tahitians. Maugham felt Brooke wrote his very best poetry during that period in the South Pacific. "The South Seas have got into my blood," Brooke wrote home, and finally as he started on his return he wrote: "Do not go beyond civilization, it is unsettling. Inside civilization one can realize the beastliness of it and labor—if one is as honest as I hope you'll be—to smash it. But when you get outside you realize the advantage of not being in it too acutely."

130

He returned to the rain of England, where the memory of the highly colored fish of the Coral Seas now turned into a sheaf of poems.

Perhaps James Michener summed up as neatly as any some of the call of the South Seas. The South Pacific, he writes, was once the playground for sick European sailors. Then it became the roistering barricade of the last great pirates. Next it was the longed-for escape from the canyons of New York, then the unwilling theater of an American military triumph, but now it has become the meeting place for Asia and America.

As such, perhaps, its story has just begun.

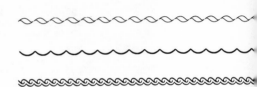

8

Society's Islands

"THEY'RE ALL GONE," the old man said; "it's the end of the season, you know. Oh, there may be one or two around; some like to stay and see all that beautiful weather we have in early May—the hibiscus blooming like crazy and the sea grape just as shiny as the gill of a mullet. For myself, I don't know why they ever leave it this time of year. Late April in Florida the fishing is good; of course, they don't fish. Swimmin's not so hot, all those jellyfish, but just take a look at that sky. Who'd want to give that sky up? Why, here on the island, we've got the best sky in the world."

He stood guarding the bridge. He wasn't the regular tender, but he was old and garrulous in the sun and what he had to say was not really about the people, just about the land, this beautiful Jupiter Island. Curiously enough, it is loved by two completely different groups of people: the local Crackers who have known it throughout their lives and in the memories of their parents and their parents before them; and "society," a handful of people who love it in part for the same reason—it is one of the most beautiful of all islands on our coast. To the old man it is the island; to society it is Hobe Sound.

It's an old saying that the coasts of Florida, from the head of Indian River on the east to Tampa Bay on the west, are one of the world's great cruising grounds. So many throughout the years have known Jupiter,

132

Plum Orchard. This gracious mansion, one of the several residences of the Carnegie heirs on Cumberland Island, Georgia, seems to epitomize the enchantment of an island retreat. (Courtesy National Park Service)

particularly years ago when the Jupiter Light was one of the great sources of comfort as one scanned the distant sky for a mooring. The cruising pattern for the old skipjacks of Indian River fame used to start at Jacksonville, making their way from island to island.

Those old Florida skipjacks were about twenty-nine feet overall with a thirteen-foot beam. There was a booby hatch over the cabin, and the cabin itself, about thirteen feet by ten feet wide, was divided fore and aft by a centerboard trunk. Those were the native boats, far different from the ones now used by the residents of Hobe Sound. The Jupiter Inlet on the coast of Florida knew them well. On the south end of Jupiter the old sailors of the last century used to make camp. They would be welcomed by old Captain Carlin in command of the lifesaving station at Jupiter. It was he who used to give advice about how to sail out with the strong flood running; how to get across the inlet and across the bar without shipping water; where to find the shark and the sawfish; how to shape a course south running parallel to the beach, keeping a quarter mile outside of the surf to avoid the strange current of the Gulf Stream,

Thousands of visitors to Nassau have viewed this $25,000 silver bar taken from a Spanish galleon off Great Abaco Island in the Bahamas in January, 1952. Markings indicate the bar was intended for Philip, King of Spain, in the seventeenth century. (Courtesy Bahamas News Bureau)

which can dash bitterly against the coast here. In those days the society of the islands was the fish and the animals—pelicans sitting on the mangrove trees, the raccoons and possums in the sand. But now if one listens to the stories of Hobe Sound, they are of other flora and fauna— names such as Rockefeller and Weyerhaeuser, names that make history, that make money; they're here on this island but almost invisible. Instead the houses are named—and a strange collection they are—for the Fords, the Armours, the Strawbridges, the Pryors, the Whitneys; they have islanded themselves beyond the sea grape.

Society likes islands; some they have developed for many years: Hobe Sound and Fishers Island, Shelter Island, areas of Long Island, Jekyll Island, sometimes private islands. There are some 8000 private islands in the United States.

The society of Hobe Sound, although old society, is new to the island itself, but the island has always had a gracious dignity about it. The Australian pines, the foliage, its proper rightness of length and breadth make it the true aristocrat, nature's rather than society's world.

ISLANDS

Discovered, they say, by Ponce de León and once sparsely inhabited by the Gilbi Indians, it now harbors fewer than 300 houses, all kinds, but all luxurious. Locally, however, they credit Ponce de León with less and Mrs. Robert Lovett with more, because she was the one who supposedly while looking on a map discovered this one island which seemed far superior to what also had once been an island, but is now a state of mind—Palm Beach. If Palm Beach did not exist, it would have to be invented, say many.

The island of Jupiter and those that live on it are subject to an unusual amount of bitterness. In Palm Beach they say everybody in Hobe Sound is a snob. In Hobe Sound they say there is nothing in Palm Beach except the shopping. In the local legends of the islands there is less talk of Ponce de León than there is of the great and deliciously wicked sayings about the sound itself: how it was developed, for example, by Greenwich's Joseph Vernor Reed, who said, "I'd like to have a cottage at Newport, but it's such a bad address." Or how novelist John P. Marquand's daughter as a young schoolgirl asked her father, "Daddy, are you the richest man in Newbury? [Newbury, Massachusetts, was her home.] Everybody at school says so." But when they vacationed in their cottage on Hobe Sound, sandwichesd between Morton Salt and Du Pont, she asked, "Are you the poorest man in Hobe Sound?" Or how the Duchess of Windsor is permitted to come up for dancing, but they wouldn't want her to live there. Or how Jacqueline Onassis's mother once tried to buy into the community and could not. Or how it is said that "they now live in Hobe Sound because Jekyll Island turned into Coney Island."

Jekyll Island was one of the first of the truly social islands. It is one of those beautiful chains of sea islands of magnolia and bay, live oak and cypress, moss and scrub palmetto that have long known devotees. In the chain there are seven large coastal islands: Cumberland, Jekyll, St. Simon's, Sapelo, St. Catherine's, Ossabaw, and Skidaway, or Tybee. Once they were the home of Jesuit missions, where the fathers learned to catch shrimp, crab, and turtle, to hunt wild turkey and duck. On these islands the Franciscans planted pomegranates and oranges and began to know the traders. A French fleet soon came to take from the Indians the wild deer, the beaverskins, and the sassafras. There followed then a period of disorder throughout all the islands and eventually they were all abandoned. When Jonathan Dickinson late in the seventeenth century made his famous trip in a canoe after a shipwreck on the Florida coast he said, "The islands are abandoned and uninhabited, but show traces of having once been posts of great importance."

It was General James Oglethorpe who was responsible for the settlement of Jekyll Island. He planted rows of orange trees, some ten thousand of them, and planted, too, the hops and grain for the brewery that was to be established. Jekyll remained a military reservation until 1766, but soon all of the islands began to change. Grants were given. St. Simon's for example, became a Scottish enclave with MacKays and MacIntoshes, Cuthberts and Grants. Great homes were built, some by the Huguenots, who made elaborate plantations, "conspicuous for beauty and comfort." William Bartram in 1774 made a trip collecting flowers and shrubs throughout this area:

In the month of March, 1774, I set off from Savanna, for Florida, proceeding by land to the Alatamaha, where I diverted my time agreeably in short excursions, picking up curiosities, until the arrival of a small vessel at Frederica, from Savanna, which was destined to an Indian trading house high up St. John's, in East Florida. Upon information of this vessel's arrival, I immediately took boat and descended the Alatamaha, calling by the way of Broughton island, where I was kindly received by Mr. James Bailey, Mr. Lauren's agent. Leaving Broughton island in the evening, I continued descending the south channel nine or ten miles, when, after crossing the sound, I arrived at Frederica, on the island of St. Simon, where I was well received and entertained by James Spalding, esq. This gentleman, carrying on a very considerable trade, and having extensive connections with the Indian tribes of East Florida, gave me letters to his agents residing at his trading houses, ordering them to furnish me with horses, guides, and every other convenient assistance.

135

Before the vessel was ready to sail again for St. John's, I had time to explore the island. In the cool of the morning early, I rode out of the town, directing my course to the south end of the island. After penetrating a thick grove of oaks, which almost surrounded the town on the land-side, suddenly a very extensive and beautiful green savanna opened to view, in length nearly two miles, and in breadth near a mile, well stocked with horned cattle, horses, sheep, and deer. Following an old highway, now out of repair, across the savanna, I ascended the sloping green bank, and entered a noble forest of lofty pines, and then a venerable grove of Live Oaks, under whose shady spreading boughs opened a spacious avenue, leading to the former seat of general Oglethorpe, but now the property of capt. Raimond Demere. After leaving this town, I was led into a high pine forest; the trees were tall, and generally of the species called Broom-pine (P. palustris Linn.) the surface of the ground covered with grass, herbage, and some shrubbery: I continued through this forest nearly in a direct line towards the sea coast, five or six miles, when the land became uneven, with ridges of sand-hills, mixed with sea-shells, and covered by almost impenetrable thickets, consisting of Live Oaks, Sweet-bay (L. Borbonia),

136

This encrusted cannon lay unnoticed on the bottom of Nassau harbor for over a century until expert eyes identified the ancient artifact for what it is. Relics like these have led to discovery of valuable treasure in the Bahamas. (Courtesy Bahamas News Bureau)

Myrica, Ilex aquifolium, Rhamnus frangula, Caffine, Sideroxylon, Ptelea, Halesia, Callicarpa, Carpinus, entangled with Smilax pseudo-china, and other species, Bignonia sempervirens, B. Crucigera, Rhamnus volubilis, &c. This dark labyrinth is succeeded by a great extent of salt plains, beyond which the boundless ocean is seen. Betwixt the dark forest and the salt plains, I crossed a rivulet of fresh water, where I sat down a while to rest myself, under the shadow of sweet Bays and Oaks; the lively breezes were perfumed by the fragrant breath of the supérb Crinum, called by the inhabitants, White Lily. This admirable beauty of the sea-coast-islands dwells in the humid shady groves, where the soil is made fertile and mellow by the admixture of sea shells. The delicate structure of its spadix, its green broad leaves, and the texture and whiteness of its flowers, at once charmed me. The Euphorbia picta, Salvia coccinea, and Ipomea erecta, were also seated in front of my resting place, as well as the Lycium salsum (perhaps L. Asrum Linn.) a very beautiful ever-green shrub, its cerulean flowers, and coral red berries, always on its branches, forming not the least of its beauties.

Time now admonishing me to rise and be going, I, with reluctance, broke away from this assembly of maritime beauties.

Continuing on, southward, the salt plains on my left hand insensibly became narrower, and I at length reached the strand, which was level, firm, and paved with shells, and afforded me a grand view of the boundless ocean. : . .

So this was the Eden of Georgia, as it was called—islands that Fanny Kemble loved, that Aaron Burr wrote about, that Sidney Lanier sang about. But these sea islands, these islands of cotton had more than one viper in their bosom: They had those "mercantile interests." Fanny Kemble looked upon the slavery in the area and found it wicked. The northerners might have called Fanny Kemble beautiful, but to many of the southerners she was simply an ugly little dark woman and when she went to one of those family plantation islands, Butler's Island, as a wife she cried out against the abomination of desolation, "the black history of oppression."

In 1886 after a long period of desolation in the islands, Jekyll Island was sold by the descendants of the du Bignon family to a group of New York businessmen that included the elder Morgan, George Baron, James Hill, and William Vanderbilt. It became the island of "the one hundred

A view of Capri and the Faraglioni. (Courtesy Italian State Tourist Office)

millionaires" because it was a club limited to that many. Here from January to March, a somewhat shorter season than that to develop in Florida, society gathered, and from here in one fell swoop society departed. Nobody quite knows why Jekyll fell so completely. There were those who said that at one time on this one island were men who controlled one sixth of the world's wealth, but by June, 1947, the state of Georgia had condemned the island and for a bit more than half a million dollars turned it into a State Beach Park. Perhaps more than anything else Jekyll fell because the age of the club fell.

Far south of Jekyll Island there is a dream built on sand—as a matter of fact, a series of dreams, areas that were developed by Henry M. Flagler, the greatest dreamer Florida was ever to know. From Jacksonville to the tip of Key West his name became a legend, but his personality perhaps was never more dominant nor more aggressive nor more attacked than it was in that island paradise—that new Eden of the United States as it was called—Palm Beach.

Florida after the Civil War began to know its first tourists. Soldiers who had seen it during the war returned and settled. In October, 1867, a man called George Seers cruised along the Florida coast. When he saw a handsome inlet he sailed through a shallow pass and entered into a glorious tropical lake. It was a beautiful sight. The rich jungle vines hung over the side of the water reflecting in the late afternoon sun the shape of sea grape, coconuts and all the rich beauty of a Florida fall. This was Lake Worth, named for Gen. William J. Worth by soldiers who had settled Indian troubles in that section of the country years before. Now Seers found no one on the barrier island east of Lake Worth, truly a deserted paradise, except for two deserters from the Confederate Army. There they lived in quiet isolation unaware that the war had been over for more than two years.

The news of this new haven began to spread northward. *Harper's Magazine,* that mentor of the public imagination, in an issue of 1870 told the story of this new Eden to the world. Here it was summer all year long. Here you could fish and put fish into your boats in such a way that they almost sank below the waterline and could barely move. Here Indians came to you with the freshest of venison, spread it on the ground for you to choose your favorite portion. Here you could get a turtle weighing 500 pounds, and best of all, so the story said, you could get land for five cents an acre.

Naturally these dreams of the South were in part fantasy, but nonetheless the new resorters began to come even then, including a few great names like Robert R. McCormick, for example, who built a winter home on Lake Worth, Capt. O. S. Porter, and others. Lake Worth's name by

1886 had been changed to Palm Beach supposedly because of the coco-nuts that had been washed up in a Spanish wreck years before and had produced palm trees.

Flagler did not discover it until 1893. A good publicist, he spread the word around. Palm Beach was quite literally on the map. On May 1, 1893, he started work on the Royal Poinciana Hotel. The building materials were sent to Jupiter and then transported down the line on a little railroad called the Celestial Line. Across the way in West Palm Beach were the shanties or the tents for those who would be working on the hotel. The workmen rowed across the tropical lake each morning and back each afternoon. Actually West Palm Beach grew faster than its more exotic neighbor to the east. But as soon as the Royal Poinciana was finished the guests were legion. It was one of the largest buildings in the world to be used for a hotel, and Flagler used to love to brag about the materials used in its making: 5,000,000 feet of lumber, 360,000 shingles, 4000 barrels of lime, 1400 kegs of nails, 240,000 gallons of paint and then that figure—impressive for the time—of $1,000,000 of building money. With 540 bedrooms, with its veranda, lounge, parlors, casino, and its green velvet chairs it soon became one of the famous hotels of the era. It added its contribution to the Gay Nineties with a richness and gaiety that many were never to forget.

139

Flagler followed the Poinciana with the Breakers, first known as Palm Beach Inn. The Breakers, having burned several times, was rebuilt in 1925 and is still considered one of the great hotels in the United States. In 1913 Carl Fisher built the bridge that started the development of Miami Beach, also an island, seventy miles to the south.

In 1918 Addison Mizner made the scene, quite literally, designing and decorating the buildings that were to make Palm Beach what it is today—Whitehall (now a museum), for example, which Flagler built for his second wife, Mary, for $2,500,000, a sweet gift of a castle set in Australian pines, palms and flowering shrubbery. The interior was spectacular: Bronze doors led to the grand hall; the hall was over a hundred and ten feet long and forty feet wide and—naturally—of white marble, worked further with black marble. It had rococo touches of course, including a clock nine feet high, called Representing Time. The Hearst of his day, Flagler furnished the place in a variety of styles with everything from antique Florentine chests to Spanish tapestries. There was Aubusson tapestry made for each chair, the windows were rich with Colbert lace and in the library there was a massive picture of Flagler's competitor, Ponce de León, who discovered Florida just a little before Flagler. The pipe organ was the largest in the United States. The ballroom was the most nearly perfect since the time of Louis XV. The guest rooms were

Jupiter Inlet, Florida.

Camp at Jupiter Inlet, Florida.

a mass of color and history, each one representing a different country—Spain, England, France, Italy, sixteen of them, with sixteen private hallways leading to the main hall.

But what about the North? Where was the "World's Most Fashionable Island"? Buried now in the democracy of the Long Island Expressway, it needs an archaeologist to reconstruct it . . . or the old files of *Fortune*, that great vault for past mores and ancient glories.

It was indeed called the World's Most Fashionable Island in 1932, and it is a little difficult now to realize that the description referred to Long Island. A *Fortune* map of that year showed only a few points of interest. There was first Queensboro Bridge, a portal even then to the motorists' inferno. The Queensboro then led to "a paradise" as it was described—the smart and wealthy North Shore, the eminent Hamptons, the Westbury-Hicksville polo district. "Little else," they said in those days, "mattered."

After the Queensboro Bridge the important spot was the Morgan estate. Here on West Island, shared by Junius Spencer Morgan, Jr., and Mrs. William L. Harkness, was where the house of Morgan celebrated Christmas; here on East Island was J. P. Morgan's own domain. The gardens at Matinocock were famous and so was the Morgan story because in 1931 Morgan had won a suit to have his assessment of $2,256,000 cut a million dollars. In the same vicinity was Lattingtown with its three square miles assessed at $7,000,000. Incorporating, *Fortune* commented, a handful of neighboring estates into a "village" is a rich man's racket" along the North Shore. But it was nonetheless a nice place to live and still is, although times have changed. In 1932 the Creek Club had as its president Clarence H. Mackay, and the members included J. P. Morgan, George F. Baker, Thomas W. Lamont, Artemus L. Gates, Edward F. Hutton, F. Trubee Davison, S. Parker Gilbert, and Harvey D. Gibson, whose estate bordered on that delicious location, the Creek Club itself.

It was an island where Belmont was "the premier U. S. race track," where there were social citadels of the South Shore, a land as *Fortune* explained with estates that were bounded by "trees, Italians and millionaires" until one reached the jumping off place of Montauk Point.

Beyond Montauk Point, still belonging to the island, was Fisher's Island and nestled between the flukes of the island was Shelter Island. Offshore was the private kingdom of Gardiner's Island.

If Florida's islands offered sun, fun, and games, and Long Island horses and haute cuisine, Mt. Desert in Maine offered the rockbound intellect an opportunity to vacation. Here a different society wandered over the rocks—the aristocractic New England mind on holiday. Oliver Wendell Holmes mused about that aristocracy:

141

ISLANDS

We are forming an aristocracy, as you may observe, in this country,—not a gratia-Dei, nor a jure-divino one,—but a de-facto upper stratum of being, which floats over the turbid waves of common life like the iridescent film you may have seen spreading over the water about our wharves,—very splendid, though its origin may have been tar, tallow, train-oil, or other such unctuous commodities. I say, then, we are forming an aristocracy; and, transitory as its individual life often is, it maintains itself tolerably, as a whole. Of course money is its corner-stone. But now observe this. Money kept for two or three generations transforms a race,—I don't mean merely in manners and hereditary culture, but in blood and bone. Money buys air and sunshine, in which children grow up more kindly, of course, than in close, back streets; it buys country places to give them happy and healthy summers, good nursing, good doctoring, and the best cuts of beef and mutton. When the spring-chickens come to market—I beg your pardon,—that is not what I was going to speak of. As the young females of each successive season come on, the finest specimens among them, other things being equal, are apt to attract those who can afford the expensive luxury of beauty. The physical character of the next generation rises in consequence. It is plain that certain families have in this way acquired an elevated type of face and figure, and that in a small circle of city-connections one may sometimes find models of both sexes which one of the rural counties would find it hard to match from all its townships put together. Because there is a good deal of running down, of degeneration and waste of life, among the richer classes, you must not overlook the equally obvious fact I have just spoken of,—which in one or two generations more will be, I think, much more patent than just now.

You could find them there on Mt. Desert—not collecting scandalous stories to repeat at dinner, but down on their hands and knees collecting marine life in Bar Harbor. At the turn of the century Augusta Foote Arnold gave us some directions that are still valuable today:

The beautiful coast of Maine is a particularly good field for shore-collecting. The rocky coast harbors the boreal fauna and flora which depend upon such physical conditions, and the shores at Bar Harbor are typical of those found elsewhere in northern New England. The rocks give shelter from the beating surf, while life has exposure to the cold, pure waters of the arctic current. Everywhere along the shore, rock pools are to be found. These are perhaps the most fascinating of all spots to the collector. They are veritable gardens of the sea, where species flourish which naturally belong to deeper water, but which find in such pools conditions suitable to their existence.

At Bar Harbor one well-known and frequently visited rock pool is found in Anemone Cave. Entering a field at Schooner Head, one turns to the right and follows the rocky shore for two or three hundred feet. It is difficult to take this short walk without being constantly diverted and de-

View at Cape Island City, Cape May Island, N.J. (1865). This old engraving shows the Atlantic Hotel and Columbia House.

143

Castle Head, Bar Harbor, Maine.

layed by the various attractions one meets, such as the tide-pools, the barnacles which in places whiten the rocks, the periwinkles, the purpura shells, and the curious algae; but at last one arrives at a cavern under an overhanging rock. Here is a large tide-pool which at first sight displays only a beautiful scheme of color. It is carpeted with a bright-pink alga, Hildenbrandtia rosea, which incrusts the basin of the pool.

Interspersed with the pink are patches of a deep-red color, having a velvety appearance, which are formed by another crustaceous alga, Petrocelis cruenta. The water of the pool is of crystal clearness, and as one gazes into it one object after another comes into view, until one is filled with astonishment at the number of beautiful objects the pool contains. The little green balls, one half of an inch to one inch in diameter, which look like small green tomatoes scattered on the stones, are Leathesia difformis, an alga which cannot be mistaken for any other. Bunches of Corallina officinalis, which resembles coral, as the name indicates, are abundant. This alga should be examined with a magnifying-glass. It is covered with calcareous matter, and its peculiar form of growth is beautiful and interesting.

The fronds of the laminarian Alaria esculenta are tiny here, while just outside the cave they are to be seen several feet in length, beating against the rocks in the swash of the waves. Thorny sea-urchins (Strongylocentrotus drobachiensis) make green spots which look like tufts of moss. Yellow and green sponges in little cones are spread over small surfaces. Starfishes and ophiurans are plentiful. The Purpura lapillus and Littorina litorea and rudis (periwinkles), so plentiful on this coast, are present. The Mytillus and the Saxicava and the Acmaea testudinalis are also to be found. A green crab (Carcinus maenas) is snugly hidden in a dark nook on the shady side of the pool, and many small crustaceans scuttle away from under stones as they are lifted. The collector is always anxious for uncommon, or rather less plentiful, species, and here are found two specimens of nudibranchs, or naked mollusks, Aeolis and Dendronotus. The Chiton ruber, a jointed mollusk, was also found here, and five species of sea-anemones were counted. As this is a favorite hunting-ground, the anemones have not been left to attain full growth; but there are very many small ones which at first are not distinguishable, as they retract their tentacles at the slightest disturbance of the water and are then quite inconspicuous. After a little time of quiet watching they will be seen putting out their tentacles and expanding their beautiful flower-like forms. It is useless to try to capture them uninjured, so tightly do they adhere to the rocks, and the difficulty of preserving them in an expanded form is so great that amateur collectors had better leave them undisturbed to beautify the pool.

It was hard to resist robbing this rock pool, where the author in half an hour counted twenty different species, and finally left, feeling that its treasures were not half discovered; but collecting should be done elsewhere, and this pool be guarded as a gem to be admired and not to be despoiled.

144

This pool in Anemone Cave, although so very attractive, is surpassed in beauty and interest by pools on Porcupine Island, at the base of the cliff. This place is somewhat difficult of access, and the timid will not undertake the descent to it; but the enthusiastic collector, who overlooks small obstacles, will be repaid by a visit to this spot, where all the treasures of Anemone Cave are multiplied many times over. These pools are resplendent with large anemones, hydroids, nudibranchs, mollusks, echinoderms, crustaceans, and algae. Alaria esculenta, several feet in length is beaten to a fringe against the rocks, and Agarum Turneri, the sea-colander, is also found here, together with beautiful specimens of Rhodymenia palmata, which is so plentiful that it reddens the rocks. . . .

Presidents of universities including Eliot of Harvard, bishops and artists, writers and, eventually, billionaires (both John D. Rockefeller, Jr., and Edsel Ford spent large amounts developing Mt. Desert) all comfortably ensconced themselves in the spot that Dr. S. Weir Mitchell had described as "a jolly place—mountains in the middle; eight lakes; sea water all around island, sometimes five foot high, sometimes 800. Lots of caves full of pools; pools full of anemones. I'd rather see it than a circus, wouldn't you?"

When island kingdoms are mentioned, one thinks of England, for example, when Elizabeth I made it *her* particular kingdom, with more than hereditary right, but also by the right of peculiar, personal possession; or the islands around Scotland; or an anachronism today, the Dame of Sark's island of Sark; or that colorful island of Sumatra ruled by Sir Stafford Raffles that was beautifully satirized in the style of popular geographical writing of the nineteenth century by Oliver Wendell Holmes:

145

This island is now the property of the Stamford family, having been won, it is said, in a raffle, by Sir _____ Stamford, during stock-gambling mania of the South-Sea Scheme. The history of this gentleman may be found in an interesting series of questions (unfortunately not yet answered) contained in the "Notes and Queries." This island is entirely surrounded by the ocean, which here contains a large amount of saline substance, crystallizing in cubes remarkable for their symmetry, and frequently displays on its surface, during calm weather, the rainbow tints of the celebrated South-Sea bubbles. The summers are oppressively hot, and the winters very probably cold; but this fact cannot be ascertained precisely, as, for some peculiar reason, the mercury in these latitudes never shrinks, as in more northern regions, and thus the thermometer is rendered useless in winter.

The principal vegetable productions of the island are the pepper tree and the bread-fruit tree. Pepper being very abundantly produced, a benevolent society was organized in London during the last century for supplying

"Norwood," historic Bermuda home in Hamilton, is more than 200 years old and overlooks the harbor. The garden contains a maze which is a miniature replica of the maze at England's Hampton Court. (Courtesy Bermuda News Bureau)

146

Cumberland Island, Georgia, has been recommended for preservation as a national park. (Courtesy National Park Service)

the natives with vinegar and oysters as an addition to that delightful condiment. (Note received from Dr. D. P.) It is said, however, that as the oysters were of the kind called natives in England, the natives of Sumatra, in obedience to a natural instinct, refused to touch them, and confined themselves entirely to the crew of the vessel in which they were brought over. This information was received from one of the oldest inhabitants, a native himself, and exceedingly fond of missionaries. He is said also to be very skilful in the cuisine peculiar to the island.

During the season of gathering the pepper, the persons employed are subject to various incommodities, the chief of which is violent and long-continued sternutation, or sneezing. Such is the vehemence of these attacks, that the unfortunate subjects of them are often driven backwards for great distances at immense speed, on the well-known principle of the aeolipile. Not being able to see where they are going, these poor creatures dash themselves to pieces against the rocks or are precipitated over the cliffs and thus many valuable lives are lost annually. As, during the whole pepper-harvest, they feed exclusively on this stimulant, they become exceedingly irritable. The smallest injury is resented with ungovernable rage. A young man suffering from the pepper-fever, as it is called, cudgelled another most severely for appropriating a superannuated relative of trifling value, and was only pacified by having a present made him of a pig of that peculiar species of swine called the Peccavi by the Catholic Jews, who, it is well known, abstain from swine's flesh in imitation of the Mahometan Buddhists.

147

The bread-tree grows abundantly. Its branches are well known to Europe and America under the familiar name of macaroni. The smaller twigs are called vermicelli. They have a decided animal flavor, as may be observed in the soups containing them. Macaroni, being tubular, is the favorite habitat of a very dangerous insect, which is rendered peculiarly ferocious by being boiled. The government of the island, therefore, never allows a stick of it to be exported without being accompanied by a piston with which its cavity may at any time be thoroughly swept out. These are commonly lost or stolen before the macaroni arrives among us. It therefore always contains many of these insects, which, however, generally die of old age in the shops, so that accidents from this source are comparatively rare.

The fruit of the bread-tree consists principally of hot rolls. The buttered-muffin variety is supposed to be a hybrid with the cocoa-nut palm, the cream found on the milk of the cocoa-nut exuding from the hybrid in the shape of butter, just as the ripe fruit is splitting, so as to fit it for the tea-table, where it is commonly served up with cold——

Colorful and apart, these islands and their rulers are alien to our culture, it would seem, but the truth is our coastlines were once girdled with private islands in the aristocracy of the sea.

In the very first issue of *Fortune* (February, 1930) sandwiched between articles on Banks and Glass, Island Kingdoms held their own.

Rainsford's Island, Boston Harbor, in a painting by the American artist Robert Salmon. (Courtesy Museum of Fine Arts, Boston, Karolik Collection)

148

"Behold the 20th century refugee. Offshore he builds his home, makes his laws, flies if he will, his own flag."

There was, for example, the island kingdom of the chocolate king of France, Henri Menier, who gave himself a private sweetmeat, his own country, when he bought Anticosti in Canada, that island of strange mists and terrible shipwrecks. A kingdom more than half the size of Connecticut, it had the isolation, and the space, for Menier—space for the mansion he would erect, every plank screwed into place; space for the hunting lodge he envisioned. But he had not lost the instincts of a trade baron. Surrounded by his objets d'art and with his Aubusson tapestries to keep him warm, he decided to forget the lodge and build instead a pulp mill. Hunting leases were granted to Americans, but it was up to them to find what creature comforts they could. If they were there, let them hunt for lodgings as well, perhaps with some of the thousand farmers Menier had brought from Europe to develop silver-fox ranches.

Menier had paid $175,000 for the island, a sweet sweetmeat indeed, and on his death it was sold to the Wayazamack Pulp and Paper Co., Ltd., for $6,000,000.

Heart Island, in the Thousand Islands, has a more familiar story. Those who had a fondness for the old Waldorf Astoria Hotel can get a bit of nostalgia as they pass its lonely sister, the castle on Heart Island.

Resting near the "long walk." (Nineteenth-century print from the authors' collection)

149

With walls of Oak Island granite, with a Romanesque powerhouse, it was an architectural horror and wonder. No wonder, of course, that it looked like the Waldorf: George Boldt, a great hotelman, owned both. He could not change the shape of the concrete heart of Manhattan, but damn it, he could change the shape of his own island. If nature had no heart, Boldt did—and the island was carved and gutted until it resembled one. But it became an island of a broken heart. At the death of his wife, Boldt abandoned the island—the pond, the swan lakes.

The island kingdoms of Maine were naturally more restrained. On Chanterelle Island, Mrs. May Harris of Bala, Pennsylvania, needed no pretension. Twelve cottages, three boats, and a recreation hall were sufficient for a refuge from the Main Line.

They were even more rustic in Boothbay Harbor, where G. Ellsworth Huggins and Harry Emerson Fosdick bought Mouse Island. Fosdick explained its attraction: "It is far enough from the mainland so that we can live an entirely unsophisticated life; that is to say, a man can put on a flannel shirt in the morning and go to bed in it at night if he feels like it."

Four miles from Bar Harbor, Ironbound Island had some ironbound prohibitions instituted by its owner, Dwight Blaney of Beacon Hill, Boston. The only form of illumination allowed was candlelight—occasionally irritating John Singer Sargent, who worked on many Maine landscapes from the vantage point of Ironbound.

Bartlett's Island, also in Maine, was already in 1930 what many private islands have become in our own time—wild-life refuges and sanctuaries. Augustus Peabody Loring and William Caleb Loring allowed no shotguns on *their* island and deer, fox, mink, rabbit, muskrat, black duck and partridge abounded.

The Isles of Shoals had a kingdom whose size came and went with the tide. Lunging Island, owned by the Reverend F. B. Crandall, was five acres at low tide, three at high tide—but it was always large enough to contain a cottage for the needs of any shipwrecked mariner—the doors never locked, a fire always laid.

Massachusetts' kingdoms were as individual as their owners. On Joshua Crane's island peacocks strutted with imported Hungarian pheasant. "I have just been elected," wrote Crane to his classmates, Harvard '90, "King of No Man's Land, southernmost island and piece of land in Massachusetts."

Then there was one of the most famous private islands of all, Naushon—*The* island, some called it—none better, again, than Oliver Wendell Holmes:

Where have I been for the last three or four days? Down at the Island, deer-shooting.—How many did I bag? I brought home one buck shot.— The Island is where? No matter. It is the most splendid domain that any man looks upon in these latitudes. Blue sea around it, and running up into its heart, so that the little boat slumbers like a baby in lap, while the tall ships are stripping naked to fight the hurricane outside, and storm-stay-sails banging and flying in ribbons. Trees, in stretches of miles; beeches, oaks, most numerous;— many of them hung with moss, looking like bearded Druids; some coiled in the clasp of huge, dark-stemmed grape-vines. Open patches where the sun gets in and goes to sleep, and the winds come so finely sifted that they are as soft as swan's down. Rocks scattered about,— Stonehenge-like monoliths. Fresh-water lakes; one of them, Mary's Lake, crystal-clear, full of flashing pickerel lying under the lily-pads like tigers in the jungle.

Such hospitality as that island has seen there has not been the like of in these our New England sovereignties. There is nothing in the shape of kindness and courtesy that can make life beautiful, which has not found its home in that ocean-principality. It has welcomed all who were worthy of welcome, from the pale clergyman who came to breathe the sea-air with its medicinal salt and iodine, to the great statesman who turned his back on the affairs of empire, and smoothed his Olympian forehead, and flashed his white teeth in merriment over the long table, where his wit was the keenest and his story the best.

An August day at the seashore, Long Island. (Nineteenth-century print from the authors' collection)

151

Gathering autumn leaves in Southampton. (Engraved by Henry Wolf for "St. Nicholas," from the painting by William M. Chase)

ISLANDS

Naushon Island near Martha's Vineyard was the private island of the Forbes family of Boston. It was to this island that Ralph Waldo Emerson sailed once with an extreme case of seasickness, but nonetheless wrote kindly about his host. "It is a matter of course that he should shoot well, ride well, sail well, administer railroads well, carve well, keep house well, but he was the best talker of the company. . . . I divided my admiration between the landscape of Naushon and him." He, of course, was John Murray Forbes of the Boston mercantile house, founder of railroads, gentleman and scholar who in 1857 bought Naushon Island, where for three generations other Forbeses from the director of the Fogg Art Museum to the Governor of the Philippines, have had their summer homes.

Those delightful islands of Boston Harbor, too, had a private island—Brewster Island, owned by the late Melvin Ohio Adams, who sailed to it in a dory, the *Leviathan*. Later his son Karl Adams commuted to his Tremont Street law office by powerboat. Karl Adams maintained that the island was a republic—the Federal Government had seized it once during World War I, released it in peace, but the Commonwealth of Massachusetts had no jurisdiction and collected no taxes. A twelve-acre rock, it is a green island in the harbor lighted by the famous Boston light and blasted at one time by its fog signal.

Harvard, an omnipresent landowner, owned Bumpkin Island, also in Boston Harbor, leased to Alfred C. Burrage for ninety-nine years. At one point it was a hospital for crippled children; at another time it was a mooring for the 260-foot yacht, *The Aztec*. It was Burrage who had organized the Chile Copper Company and was a partner of the Guggenheims.

In the *Fortune* study the editors rejoiced that "the oldest piece of property in America in the possession of the original owning family" should naturally be an island. In 1635 the Crown granted an island in the James River to Frances Eppes, a member of the Colonial Council. At the time of the study it was still owned by Mary Eppes, his descendant.

But Assateague, owned in the thirties by Samuel B. Field of Baltimore, has changed perhaps most of all. An island of oysters, an island of the once belligerent Assateague Indians, it is an island of marshes, lagoons, salt creeks, wild fowl, and the famous Chincoteague ponies. When it was Field's private island, he had over 200 saddle ponies and herds of Hereford and Angus cattle. Now, of course, most of the island is a National Seashore preserve.

Bogue Island in North Carolina, owned by Mrs. A. E. Hoffman, was described nearly forty years ago as a place where every tree was protected by a lightning rod and where any picnicker would be shot at by an armed guard. But Mrs. Hoffman walked Bogue Island, disturbed because at the end of the seventeen-mile-long private island was Salter Path, where a hundred and seventy-one people, deep-sea fishermen and their families, lived in primitive communism long before the Hoffmans. They were tenacious squatters like moss on the trees.

If *Fortune* gently satirized Mrs. Hoffman, it admired Joseph Palmer Knapp, "The Squire of Knotts Island" of Currituck County, North Carolina, who bought his island first as a hunting lodge, only to fall in love with the people and the place.

Sapelo Island was owned by Howard E. Coffin of Detroit, who had as his guest Calvin Coolidge, and the nation focused its interest on this little-known spot. As we'll discover, it was later offered for sale when affluence and Coolidge disappeared from the American scene.

North of Sapelo, St. Catherine's Island was bought by C. M. Keys of Curtiss Aviation. Claimed in 1566 by Menéndez for the King of Spain, it was another of the islands which attracted the Jesuit missionaries. It was sold to Button Gwinnett of Declaration of Independence fame by a halfbreed Indian woman, Mary Bosomworth. In 1930 *Fortune* was still extolling it in words stranger than the old ads that interspersed the articles:

153

Surrounded by cotton fields stand the old slave quarters, square barrack blocks that face each other across the "street." Old southern darkies who think in terms of the boss and the children still live in the quarters that once sheltered their slave fathers and grandfathers.

Florida, a land of islands, of caves and grottos, lagoons, savannahs, and coral beaches, knows many islands, but the most famous one in predepression, predevelopment days was Treasure Island, owned by John T. McCutcheon, the famous cartoonist of the *Chicago Tribune*, who needed some respite from the Bible Belt and sought "a tropical island in the pirate belt."

Louisiana had Grande Isle, the home of the terrapin king, John Ludwig, and Las Conchas Island, the home of the salad-oil king, A. D. Geoghegan, of the Wesson Oil and Snowdrift Co., Inc. Even California could not be more exotic.

The coast of California had its private islands, too, in those golden days of the very early thirties before the gold was chipped away by hard times. M. F. Bramley, who used to go cruising along the Southern California coast, had his heart set on an island. He used an unusual faculty to obtain the site he most wanted; it was, he said, simply revealed to him in a dream. He acquired his island in a way that many other islands have been acquired, by the simple practice of using an old American commodity of ingenuity—and money—and built it himself. He made a circular breakwater, pumping out the unnecessary water.

William Wrigley, Jr., bought the famous island of Catalina. He found himself also buying, in that golden age, gold mines. The Wrigley mines in the early part of the thirties gave forth a stream of gold, zinc, lead, and silver. A $1,000,000 steamer carried trippers back and forth and the island itself offered the other gold of tourism. Wrigley soon built a ballroom, the largest in the country. When a foothill blocked the view he wanted, Wrigley simply had it dumped into the Pacific.

Although the government owned most of the islands, it did not own most of the rocks and Mrs. Helen K. Morton in 1925 saw a two-acre rock that attracted her attention. It was like one of those islands that spring up or that are unmarked on charts, difficult to find. Who owned it? Mrs. Morton with a woman's stubbornness tracked it down to the U.S. Land Office where they could barely admit its existence. It was, they said, in public domain. Mrs. Morton could, if she liked, homestead it, the way an old pioneer might have done. But for that she would have to live three years on the rock itself. Solitude was fine, thought Mrs. Morton, but too much of it on two acres was too much for even an "islomaniac."

154 She continued to investigate ways to obtain what she wanted, and she heard eventually of a man called Valentine who in 1846 had been assigned some 13,400 acres of land. He had not, however, properly filed his claims for the land, and he had been "squatted out" of his inheritance. But Valentine had a loud voice and was determined that he would not be beaten out by the government. He persisted; he made his presence so well known in government offices that eventually he was given certificates of scrip which allowed him 13,400 acres of public-domain property in the United States, and it did not matter where such land was. He began to be a private banker for those who wanted public-domain land. They could buy from him the necessary scrip for the land that they specifically wanted. Through Valentine, Mrs. Morton acquired her "White Rock" and owned it in glorious solitude. She was to say, "I never owned an island before. It seems even now that it may be the result of a belated reading of Cinderella and an indiscreet rarebit. I am quite frankly undecided what to do with it and am open to suggestions."

Below these Channel Islands Princess Der Ling discovered and leased from Mexico a tiny island called Golondrina off the tip of Southern California and stocked it with a strange crew. A writer who took herself back in her material to the Chinese Imperial Court, she maintained on the island her own court—guarding her palace with twenty American ex-Marines—and a hundred and fifty retainers. Rumors and stories about the princess were rampant. The Marines were under the command of

Romantic sight for Thousand Islands visitors is Boldt Castle at Alexandria Bay. (Courtesy New York State Department of Commerce)

Arthur Burks, once a lieutenant in the Marines, who supposedly had rescued the princess—a rescue worthy of the romance of her own writings—from one of the endless revolutions in China. The princess also enjoyed garbing those twenty ex-Marines in Chinese robes and having them stand around at stiff attention watching the dancers summoned for a command performance almost nightly. There the Princess of Golondrina ruled supreme, pronouncing judgments, watching the fancy drill of the Marines. Ruling the island was no easy job because the one hundred and fifty retainers consisted of a hundred and fifty women of different nationalities.

Such island splendor with such an imperial court could not, of course, last and eventually the princess went back to Los Angeles with no American Marines to guard her.

From the wild romance of Golondrina to the wild beauty of the islands off the coast of Washington is truly a step into another world. Eight miles from the town of Seattle is the island on which the Indian chief Seattle was born. It was first sighted in 1792 by Capt. George Vancouver, who gave it no name, but throughout the years it acquired its own names. At one point it was called High Island, at another time Smugglers Island, at another time Blake Island. It was finally acquired and owned by William Trimble of Seattle and was called Seattle Island. A beautiful island with many deer, it was a refuge for animal life during the hunting season. Today it is a municipal park.

But the days of the private kingdoms are not over. One of the most famous in the world is Gardiner's Island off Long Island. Gardiner's Island has been in the possession of the Gardiner family since 1639. The early maps refer to it as L'Isle du Jardinier, but the Indians gave it another name—Manchonake, the island of death. But it has always been an island of life and activity. It is, for example, an island of birds: meadowlarks and bobwhites, chimney swifts and red-winged blackbirds. Flickers abound; there are quail in the woods, indigo buntings, and enormous ospreys.

156

It is an island of greenness. At one time it harbored enormous flocks of sheep, although the island is no longer farmed. It harbored treasure too, the treasure of the legendary Captain Kidd, and the old Kidd documents still exist, one dated Boston, September 4, 1699. Most of the treasure vanished, but there is still in the family today the cloth of gold that Kidd presented to John Gardiner.

Gardiner's Island was the home of some extraordinary individuals. Lion Gardiner, who founded it, soldiered in Saybrook, Connecticut. He called himself an Englishman, but no one was quite sure about his nationality after he settled in Holland. A member of one of the English regiments under the command of Prince Frederick, he came to the attention of that band of gallants who were eager to find men to settle in America: strong men, smart men, efficient men. Lion was such a man. In Saybrook he was to build a fort and to be its commander. He grew to love the animals that abounded in Connecticut, the salt marsh that was all around him, and the color of Long Island Sound.

But the fort soon came under a siege of the Pequot Indians. Food was almost impossible to obtain and then on February 22, 1637, the Indians attacked violently, but not nearly so savagely as they would at Wethersfield, where they massacred the colonists in April.

The English turned on the Indians with almost equal savagery except for Lion, who said he would kill "only such Indians as have killed Englishmen." Peace was finally made. Saybrook became less fort and more trading post, but the soldiers began to look elsewhere, to islands perhaps where there would be peace. Having befriended the Indian Wyandance, Lion Gardiner made trips to Montauk Point and discovered a star-shaped island which he felt to be peculiarly his own. Wyandance told him it was the Island of Death, or The Island Where Many Have Died, perhaps because it had once been a spot where many Indians had died in some long-forgotten attack. First acquiring the royal grant, Lion then acquired the Indian grant—for ten bolts of trading cloth.

The island has known other names in history besides the rugged Lion and the adventuresome John. Julia Gardiner, who was called the "Tigress in the White House," married John Tyler, a man good enough for any empress of an island, the President of the United States.

Private islands are expensive to maintain, as many owners have discovered, and in addition there is enormous pressure on owners to allow their beauties to be preserved for the national good. Gardiner's may well become a wildlife refuge. Naushon may come under some control to preserve the only proved climax oak beach forest surviving in New England. The forest at the northern end of the island has never been cut in historic times, and the southern end has not been touched since 1820. Some private islands have fallen into other categories—the government putting pressures upon owners for everything from rocket ranges to more elaborate military installations.

157

But it was the depression more than anything else that changed the quality of many private island kingdoms.

Just two years later after extolling the island kingdoms, *Fortune* took note of the times and the fact that estates were going. Mowed down by the depression, empires were disappearing. For less than $1,000,000, Sapelo, the island kingdom of Mr. Coffin, was offered for sale. At that time Coffin had 1000 head of fine beef cattle, but enough pasturage for 2000. Thirty-six artesian wells supplied "the finest water in the entire country"; there were fifty miles of roads; there was the greatest live oak in the world. There were elderberries and persimmons, scuppernong grapes, wild turkey, quail, duck doves, snipes and marsh hen. The ponds were stocked with perch and bigmouthed bass, the salt streams with bass and trout.

Offshore there were tarpon, crab, and shellfish. All you needed in the depression was $1,000,000 and that kingdom could be yours. It had already fallen into slight decay and needed, as *Fortune* put it, a pretty penny to restore it to its former beauty.

Captiva Island, off the west coast of Florida, was once an area haunted by pirates. (Courtesy Florida State News Bureau)

Other island estates were going, not complete islands, but some of the great beauties of Bermuda; for example, Deepdene, available for something just over half a million dollars and one of the greatest showplaces of Bermuda. It was described in the alluring tropical type copy as a spot with one of the rarest of tropical gardens. It offered other advantages: servants for example at $3.50 a week, little or no taxes, the owner—at that point C. Ledyard Blair—paying only a parish tax (for the poor) of fifty pounds a year.

The island that had just been called The Most Social of All Islands—Long Island—was also offering up estates.

Great estates in Southampton were being threatened. Colonel Rogers's house, known to Southamptonites as The Port of Missing Men, was still sacrosanct, but his summer house with garden was up for sale. In 1912 the good colonel had bought sixty acres of seafront including two freshwater lakes. He spent $2,800,000 to turn it into what he wanted. The North Italian palazzo, described as "frankly derivative," overlooked the Atlantic instead of the island bayberry. There were arrogant hedges and geometric lanes and lawns, pergolas, loggias, glades, and statuettes—indeed thirty acres of formal garden. But even in 1932 it required $4300 a year to maintain the land. The household staff numbered eleven. It was of course just "a summer place," just a beach cottage, but it could be had for $850,000. It was the time to buy, *Fortune* cried out. The premium on U.S. estates was low, lower than ever before in the world, and if you wanted such retreats they could be had almost for the asking, almost but not quite. Society could exile itself; it cost nothing, however, if society sent you into exile.

SEAWEED: When descends on the Atlantic/The gigantic/Storm-wind of the equinox,/ Landward in his wrath he scourges/The toiling surges,/Laden with seaweed from the rocks;/From Bermuda's reefs; from edges/Of sunken ledges/In some far-off, bright Azore;/From Bahama and the dashing,/Silver-flashing/Surges of San Salvador;/Ever drifting, drifting, drifting/On the shifting/Currents of the restless main;/ Till in sheltered coves, and reaches/Of sandy beaches,/All have found repose again. —Longfellow

159

The popular "Piazza" on Italy's Capri. (Courtesy Italian State Tourist Office)

9

Islands of Exile

Dear Ussher:

   The Emperor is not very well. He wishes to delay embarking for a few hours, if you think it will be possible then. That you may not be in suspense, he begs you will leave one of your officers here, who can make a signal to your ship when it is necessary to prepare, and he will also send previous warning. I think you had better come up or send, and we can fix a signal, such as a white sheet, at the end of the street. The bearer has orders to place at your disposal a hussar and a horse whenever you wish to go up or down. Let me know your wishes by bearer. . . .

   It was no wonder Napoleon wanted to delay departure on the ship *Undaunted*, captained by Adm. Sir Thomas Ussher. Once aboard he would be deported to Elba—sent to an island, a man in exile as many were both before and after him. Napoleon's exile was fit for an emperor; only poor devils knew Devil's Island; only the wretched knew such spots as the two Australian islands of Van Diemen's Land (Tasmania) and Norfolk Island. The old missionary travelers of the nineteenth century, such as the Quaker James Backhouse, visited these islands with deep concern. They were earnest naturalists, those early missionaries, of both the soul and the land. They observed the state of both with superb clarity and they left us wonderful travel books filled with minute detail about the flora and fauna of these rare islands as well as the terrible agony suffered by many of the prisoners. The worst agony, of course, as Backhouse always used to point out, was that the prisoners themselves lost their faith; indeed there seemed to be no reason why they should have had any at all. Nonetheless, crammed more than thirty in one room, chained ankle to ankle, with the chain then secured to a tree outside, the prisoners would listen to the words of Scripture that Backhouse and his like offered as solace. "They were very attentive," he would say, "while we read to them from the Scriptures and imparted to them religious counsel; comparing the misery produced by sin, with the peace resulting from righteousness and exhorting them to flee from the former and follow after the latter."

Indeed, they had no place to flee, but the island had a strange, haunting beauty. Norfolk Island is about seven miles long and four miles wide. It is the home of that superb tree, the Norfolk Island Pine, which towers 100 feet above the rest of any forest. Backhouse observed it growing in clumps or singly on the grassy parts of the island, its roots washed in the sea at high tide. There, too, he saw the Norfolk Island Tree Fern, with the superb fronds seven to twelve feet long, digging its feet into a rain stream.

This almost inaccessible island contained the worst of the prisoners, the others being generally transported to New South Wales proper or Van Diemen's Land. On Norfolk Island, many sentenced to wear irons for the rest of their lives nonetheless worked fiercely in the limestone pits or tried to raise maize in the gardens. Often the men were imprisoned for mutiny, sailors who could not take the deprivations and cruelty of ship life, their arrogance growing stronger as they sailed deeper into the Pacific. Their defiance continued on the island. They were beaten; a hundred lashes became a small punishment. The number of lashes that each man suffered began to be a mark of bravery.

161

Sin was omnipresent; it flowered the way the Norfolk Island cabbage tree abounded. It was as stormy as the weather. "This island," wrote Backhouse, "beautiful by nature, comparable to the Garden of Eden, is rendered not only immoral with the wildness, but a place of torment to these men. No so much by the punishments of the law as by their conduct one to another. They form schemes of mischief and betray one another and being idly disposed, they are very generally chased by the exertions of the prisoner overseers to keep them at work. Being surrounded by the ocean and all of the lands being so distant the hope of escape is precluded."

But there in the beauty of Eden the men with no chance of escape would work in the sawpits of the beautiful Norfolk Pine, the trees falling around them, the wild orchids still attached to the branches, strange plants growing on the moist rocks, the ferns extraordinarily beautiful.

Johnson's Island prison during the Civil War.

Sometimes the officers of the penal settlement would urge Backhouse to go to Philip Island, an even more desolate island where one would find wild goats. There were no pigs and goats left on Norfolk by the early nineteenth century, the settlement being so far from Sydney that the animals had long since disappeared. But here on Philip Island the wild boars and the goats were accompanied by lizards and it was on all three of these that occasionally a prisoner would live if he was able to escape from Norfolk to Philip Island. Once such a man did live in the peaks of Philip Island for three months, feeding himself on the wild animals and fruit, only to feel the desolation so intensely that he gave himself up.

Often the prisoners would have been sent to some other colony first—Bermuda for example—and then charged with mutiny there, then sent on to New South Wales and finally the penal settlement on Norfolk Island.

Shortly before Backhouse's trip to the island there had been a famous mutiny in which the prisoners actually tried to take possession of the island. He now spent much time talking with men sentenced to death for that experience.

He regularly visited Van Diemen's Land, too, and there Backhouse continued his litany of both soul and soil. He was fascinated by the enormous stringy barked trees, by enormous white gums, by the great ferns that had been split as cabbage palms are today for the aborigine to obtain the food within. The great trees haunted not only Backhouse, but the prisoners. Some of them worked in forests where the trees were one hundred and fifty feet high, many of them thirty feet in circumference. "You see, sir," they would say to him, "we cannot tell, but at any hour of the day or night one of these great trees may fall upon us and

162

Early scenes on Pitcairn Island.

163

crush us; but we are prisoners sent here to work, we cannot help it." And indeed many of them were so crushed.

Backhouse was also concerned with the state of the aborigines who were often rounded up like cattle and killed. He visited Flinders Island, which had become an establishment for the aborigines. And he reflected upon what civilization would mean to the aborigine, a concern that every nation would have to consider as it distorted the traditional pattern of island peoples:

In those parts of the Colony, in which the White Population have taken possession of the lands, the Kangaroos and Emus, which were among the chief animals, on which the Blacks subsisted, have been generally destroyed, and the ground on which those animals fed, is now depastured by the flocks and herds of the usurpers of the country; who have also introduced profligate habits among the Blacks, that are rapidly wasting their race, some tribes of which have already become extinct, and others are on the verge of extermination.

It is scarcely to be supposed, that in the present day, any persons of reflection will be found, who will attempt to justify the measures adopted by the British, in taking possession of the territory of this people, who had committed no offence against our Nation; but who, being without strength to repel invaders, had their lands usurped, without an attempt at purchase by treaty, or any offer of reasonable compensation, and a class of people introduced into their country, amongst which were many, both free and bond, who, regardless of law, and in great measure exempt from its operation, by the remoteness of their situation, practised appalling cruelties upon this almost helpless race. And when any of the latter have retaliated, they have brought upon themselves the vengeance of British strength, by which, beyond a doubt, many of the unoffending have been destroyed, along with those who had ventured to return a small measure of these wrongs, upon their white oppressors.

Upon every hand, it is evident, that a heavy responsibility has thus been brought upon the British Nation; in which also, the Colonial Government is deeply involved; and that it is their bounden duty, to make all the restitution in their power, by adopting efficient measures for the benefit of the Aborigines of Australia, in affording them protection and support, and in endeavouring to civilize and settle them. . . .

I would therefore suggest, that the Government should afford the means for supplying the Blacks with food, clothing, and shelter, at all the Mission Stations; where every Black who chose to be there, at the known meal-times, which should be, at least, three times a day, should be liberally supplied with wholesome and properly cooked victuals; and that such as chose to remain for longer, or shorter periods, should be accommodated and clothed, in such a way as to give them a taste for comfort. . . .

At each of these Stations, provision ought to be made, for the board,

164

Coconuts and lava walls mark the spot on Hawaii's coast at Honaunau Bay where the ancient chiefs gave sanctuary to Hawaiian commoners.

clothing, and education of any children, that the Aborigines might be disposed to leave, for longer or shorter periods, for instruction; as it is chiefly, upon the children, that the most decided impression of civilization, may be expected to be made.

165

Encouragement should likewise be held out to other persons than missionaries, to engage in this work of benevolence, who ought to be liberally provided for by the Government, until a sufficient number of Stations should be occupied, to afford the assistance and protection, needed by the whole Aboriginal Population; who, by these means, would be drawn away from towns, and from the habitations of Settlers and Stockmen, where they are now debased and demoralized.

No work ought to be exacted from the Aborigines, for a considerable period; nor at all, except in assisting on the Establishments, at such labour as might be made to appear to them to be reasonable; but every encouragement of industry should be held out to them, by rewarding their labour, perhaps, chiefly by payments in money; in order to teach them its use and value. . . . As soon as any of the Blacks might be disposed to cultivate land for themselves, or in other ways, to adopt settled and civilized habits, they ought to receive encouragement to do so, by the allotment to them, of portions of land. . . .

Finally he collected and published a handful of pathetic letters from the "sea-girt prison" that eventually influenced the conscience of the public who demanded island prison reform:

Immigrants' hospital, Ward's Island, N.Y.

Bounty Bay and the village of Adamstown, Pitcairn Island.

Affectionate Parents,   Norfolk Island, 25th April, 1835.

An unfortunate Son, now embraces the afforded opportunity, of imparting to you, in the strongest terms, his fervent desires for your welfare, &c. Since the receipt of your last letter, I have, through the want of a friend, and which I shall long feel the need of, become convicted to Norfolk Island. . . . My present situation here however, unfortunately, for me, dear Parents, (is such) that nothing but pure conduct, and the help of God alone, can afford me again the opportunity of meeting with you in this world. . . . Dear Father and Mother, mine is a bitter lot in life; but I, myself, am alone to blame. Bid my brothers and sisters ever to bear my situation in mind, ever to refrain from drinking and loose company; and bid my sister M. to be careful over her son, lest he meet with my present unfortunate situation. . . . And may every blessing that God can impart, attend you all through life, is the prayer of your unfortunate, though affectionate Son,

Norfolk Island, 11th April, 1835.

My dear Wife and beloved Children,

Through all the chances, changes, and vicissitudes of my chequered life, I never had a task so painful to my mangled feelings, as the present one, of addressing you from this doleful spot,—my sea-girt prison, on the beach of which I stand, a monument of destruction; driven by adverse winds of fate, to the confines of black despair, and into the vortex of galling misery. I am just like a gigantic tree of the forest, which has stood many a wintry blast and stormy tempest; but now, alas! I am become a withered trunk, with all my tenderest and greenest branches lop'd off . . . to advocate; and it, ultimately, works its own cure, by destroying its own existence; and brings its supporters into that contempt, it so industriously endeavours to hurl upon that authority, to which, finally, its supporters so meanly cringe. Never permit our children to read any leading articles in any Colonial Newspapers. There is so much scurrility, vituperation, and perfidy amongst their productions. . . . I am exceedingly anxious that my dear children should have the cause of my present privations, and humiliating and degrading situation, constantly pressed upon their attention, that they never may be exposed to the same fate as that which has overtaken me, but be preserved from it. . . .

167

By the midnineteenth century Norfolk Island had been eliminated as a penal colony, but it was to continue at least for a while with a romantic history. It became associated for a while with the greatest mutiny of all times, the mutiny of the *Bounty*, when some of the mutineers' descendants left Pitcairn to try to better their lot on Norfolk. At that point Pitcairn had suffered first from a drought, then from a blight and finally from the loss of almost all the yams on which they so strongly depended. Actually some of the early descendants had migrated to Tahiti in 1831 before they tried the island of Norfolk in 1856. But when eventually a large group did go to Norfolk, they returned to Pitcairn much as did the men and women of Tristan da Cunha in our time who wanted their own volcanic island, and no place else.

Pitcairn was the island that on January 15, 1790, the group of mutineers sought for their home. But their story of course had begun long before. The mystery and the saga of Pitcairn depended, as so many historians have said, on one simple fruit, the fruit of the bread tree, that our pirate naturalist Dampier had so greatly extolled. The bread tree excited Lord Byron when he tried to construct the story of the mutiny of the *Bounty*, and in a not very good poem, he says:

> The Breadtree, which, without the ploughshare, yields
> The unreap'd harvest of unfurrow'd fields,
> And bakes its unadulterated loaves
> Without a furnace, in unpurchased groves,
> And flings off famine from its fertile breast,
> A priceless market for the gathering guest.

This remarkable breadfruit then, a kind of manna of the gods, was sought for by many early travelers, none more persistent than "Breadfruit" Bligh, a thirty-three-year-old lieutenant who in 1787 commanded an expedition to gather breadfruit and take it to the West Indies. He was in command of a ship that he described thus:

168

Her burthen was nearly 215 tons; her extreme length on deck, 90 feet ten inches; extreme breadth, 24 feet 3 inches; and height in the hold under the beams, at the main hatchway, 10 feet 3 inches. In the cock pit were the cabins of the surgeon, gunner, botanist, and clerk, with a steward room and storerooms. The between decks was divided in the following manner:— the great cabin was appropriated for the preservation of the plants and extended as far forward as the after hatchway. . . .

I had a small cabin on one side to sleep in, adjoining to the great cabin, and a place near the middle of the ship to eat in. The bulk-head of this apartment was at the after-part of the main hatchway, and on each side of it were the berths of the mates and midshipmen; between these berths the arms-chest was placed. The cabin of the master, in which was always kept the key of the arms, was opposite to mine. . . .

Unlike many of the crews of old ships, their names have come down to us. The master's mate had been specifically selected by Bligh; his name was Fletcher Christian. He was to be leader of the famous *Bounty* mutiny, together with John Adams, able seaman; William Brown, botanist's assistant; William McCoy, described in the old books as "a bad man"; Isaac Martin; John Mills, gunner's mate; Matthew Quintal, "bad man"; Edward Young, midshipman; John Williams, artisan. These were the nine men who were to land on Pitcairn to establish the colony that was to supply one of the great stories of all time.

Nonetheless many aspects of the mutiny are as fog-wrapped as most all old marine tales. Certainly, Bligh was a hard taskmaster. So were other captains contending with the length of the old voyages, the insufficient food, the fact that so many of the ships were manned by press-gangs, and flogging was omnipresent. It was an age when mutiny was in the air. Indeed, almost immediately after the *Bounty*, mutinies recurred with almost alarming frequency, as could be attested by the number of sailors who spent their last years in the Eden of Norfolk or Van Diemen's Land.

So the stage was set as the *Bounty* sailed that November day in 1787 for the Gardens of Eden in the South Seas, for the manna with which the Caribbean landmasters might feed their slaves at little or no cost. The ship was barely under way when Bligh found that there were two cheeses missing. In the old voyage literature, petty thievery seemed to be behind the most devastating consequences. Captain Cook was indirectly killed because of the hostilities arising out of the small thefts of iron by the natives, and the *Bounty*'s difficulties started with the cheese.

In the holiday season Bligh had outshone himself by allowing only two-thirds of the allowance of bread so that the ship might get to Tahiti without stopping too frequently for provisioning. Bligh, however, having picked up some pumpkins in Tenerife, decided in addition to cutting the bread to substitute pumpkin for bread, which the men patently "refused." Bligh shouted, "You damned infernal scoundrels, I'll make you eat grass or anything you can catch before I have done with you."

169

Food and their spirits began to deteriorate. Even pease and oatmeal became not only maggoty but almost nonexistent. The *Bounty* sailed on, almost everything going wrong, including its arrival in Cape Horn at the time of the worst storms. Nonetheless the ship took off again on the first of July, arriving six weeks later in Tasmania, where incidents continued to occur: men refusing to take part in small landing parties and the like. But they sailed on, finally reaching Tahiti on October 25. The crew of the Bounty became fascinated with the way of life of the Polynesians. The stay on Tahiti was too long; the breadfruit flower could not be immediately transplanted. There were bad winds and perhaps most important of all, the beauties of Tahiti were simply too attractive. As they set sail again, all nerves were taut, minor occurrences kept happening as described by James Morrison:

In the afternoon of the 27th Mr. Bligh came up, and was taking a turn about the quarter deck when he missed some of the cocoa nuts which were piled up between the guns, upon which he said they were stolen and could not go without the knowledge of the officers, who were all call'd and declared they had not seen a man touch them, to which Mr. Bligh replied

Adm. Sir Thomas Ussher, R. N., Napoleon's escort to Elba.

"then you must have taken them yourselves," and ordered Mr. Elphinstone to go and fetch evry Cocoa nut in the ship aft, which he did. He then questioned evry Officer in turn concerning the number they had bought and coming to Mr. Christian ask'd him. Mr. Christian answer'd "I do not know, sir, but I hope you don't think me so mean as to be Guilty of stealing yours?" Mr. Bligh replied "Yes, you dam'd Hound, I do. You must have stolen them from me or you could give a better account of them. God damn you, you Scoundrels, you are all thieves alike, and combine with the men to rob me.—I suppose you'll steal my yams next, but I'll sweat you for it, you rascals. I'll make half of you jump overboard before you go through Endeavour Streights." He then call'd Mr. Samuel and said "Stop these villains' grog and give them but half a pound of yams to-morrow, and if they steal them I'll reduce them to a quarter." The cocoa nuts were carried aft and he went below. The officers then got together and were heard to murmur much at such treatment. . . .

As feelings ran high Christian was the first to decide that he would mutiny, and it was with Churchill and Burkett that he finally went to Bligh's cabin and announced, "Mr. Bligh, you are my prisoner." Soon Bligh was in a twenty-three-foot launch with eighteen of his men, leaving the *Bounty* in command of Christian.

This is the Museum of Napoleon at San Martino, Portoferraio, on the isle of Elba, where the great French general went into exile following his defeat. (Courtesy Italian State Tourist Office)

171

Christian had great powers of leadership ("We shall never leave you, Mr. Christian, go where you will"). And so he was followed by the eight men—Young, Mills, Brown, Martin, McCoy, Williams, Adams, and Quintal—first to Tahiti and then on to their promised island.

On Tahiti in addition to the eight mutineers, Christian brought aboard some male and female Polynesians, one a child in arms, plants, livestock, and an enormous sense of adventure. The story from there on is filled with lacunae. After their landing on Pitcairn in 1790, the island was not rediscovered until eighteen years later. J .C. Furnas has said about it, "The best of all South Seas stories is a very famous, lurid and sentimental melodrama. It starts with institutionalized brutality, dips into psychopathology, goes heavily into sex, fights a war, proceeds with exploration in a Crusoeish vein, declines into lethal violence and finishes on a strong harmonious chord of simple piety." As the story of Pitcairn moved on, the simple piety prevailed. The descendants of the mutineers became Seventh Day Adventists as they moved into history and into the twentieth century. There are eighty-eight people left on the island according to Kenneth Bain, the British Commissioner for Pitcairn. It has a significance for Britain now because it is the closest point outside of

French territory from which French nuclear tests in the South Pacific can be monitored.

The island is one of old people.

"The future of Pitcairn," says Bain, "has a big question mark over it. The drift of the young people away from the island makes it more difficult for the elders who remain to keep the place going. There may come a time when a decision will have to be made, whether the aging population can continue to operate the island."

There is, he says, "no tourist potential whatsoever." (And how many islands can we say that about?)

Landing is hazardous: The Pitcairners are not friendly to new settlers. There is one Australian woman who has been there for thirty years and is still described to visitors as a "stranger."

Once away from Pitcairn the islanders long to return—the problem of every emigrant and once particularly poignant in our own country. . . .

They came to the New World, exiles from their past, from their villages, from their farms, from their old pain, from hunger, from persecution, from the sorry tales of their homeland. They were, of course, the American immigrants. They found themselves on islands at the first entry points to the United States. They poured into Boston Harbor, for example, only to be placed immediately on Deer Island, that once sweet island just four and a half miles due east of Long Wharf. The island, shaped like a whale, is a small one, only a mile in breadth, an island in terms of Boston history important to the old town of Boston itself and the colony of Massachusetts Bay.

Leased to various individuals, it had a quiet history until 1847, when Dr. Joseph Moriarty was put in charge of it; it was used for hospital accommodations to nurse cases of "ship fever" brought in by the immigrants and as a place of general quarantine—Boston's Ellis Island for the arriving immigrant.

The year 1847 was a painful one not only in Ireland but also in Boston. Already the streets of Boston were crowded with starving Irish immigrants. "Groups of poor wretches were to be seen in every part of the city," said the *Boston Transcript*, "resting their weary and emaciated limbs at the corners of the streets and in the doorways of both private and public houses." In ten years nearly 34,000 immigrants had landed at Boston—at least that was the legal count. There were certainly half again as many who arrived illegally. Massachusetts cried out, "Is Boston to become the moral cesspool of the civilized world? Keep them on Deer Island to see that they do not bring illness to the town." And indeed Deer Island seems to have saved the city from a terrible pestilence. Each

173

The Statue of Liberty on Liberty Island in New York Harbor has welcomed trans-atlantic travelers since 1886. (Courtesy New York State Department of Commerce)

Kalaupapa peninsula, Molokai, Hawaii, site of the Hansen's disease settlement. (Courtesy USDA)

The harbor of Portoferraio on Elba. (Courtesy Italian State Tourist Office)

174

incoming ship was inspected at Deer Island by port physicians. If there was illness aboard, the ship had to go to the south side, where it was quarantined for twenty days. From there the sick were taken to sheds and tents. Sometimes there were mistakes—one immigrant was found dead of typhus in India Street. An order for him to go to Deer Island was found in his pocket, but he could not find the boat. The next year there was an outbreak of cholera.

New York curiously enough had it easier than Boston. It was a larger city, it absorbed more people, and it had Ellis Island among other quarantine islands. The illness and pain of the immigrant did not seem so completely corrupting. Not only was there Ellis Island, there was a famous quarantine station on Staten Island, covering some thirty acres of high ground.

Those who came into New York considered themselves much better off than those who went to the quarantine stations of another island, Grosse Isle in Quebec. Those who arrived at Grosse Isle, that lovely little island of the St. Lawrence, thought they were coming to a haven, a beautiful island of rocky bays, wild flowers. One man spoke of it as "a fairy scene," with its exquisite glades, groves, wild flowers and glimpses of the St. Lawrence, but it was an island of pestilence and fever. Canada had cried out, as Boston did, that it was to be inundated with an enormous crowd of destitute immigrants. What was to happen? Information was hard to get; the authorities did not always know what ships were leaving Ireland nor when they would be likely to arrive. It was impossible to make proper arrangements.

One sweet May day on this beautiful Grosse Isle eight ships arrived with almost five hundred cases of fever. There was hardly any place to put them. Dr. Douglas, the medical officer in charge of whatever efforts were made, said, "I have not a bed to lay them on, or a place to put them. I never contemplated the possibility of every vessel arriving with fever as they do now." He was barely to get the words out when seventeen more such boats arrived. Doctors and nurses, priests, even the relatives abandoned the ill. People lay on the green grass of the island and cried for water. Children cried from hunger. When there was food, it was served and eaten half raw.

No, New York would not have that and for its time it managed the immigration problems with reasonable success. New arrivals came in to Staten Island, at that time one of the most beautiful islands of our country, some fifty-seven square miles, a summer resort of great beauty. The resort people complained of the wretchedness of conditions on the island, so much so that by 1858 there was an enormous riot that burned down the hospital. An effort was made to improve the island as well as Ward's, Bedloe's, and Blackwell's islands—all became fever hospitals for the immigrants, or convalescent homes. 175

As ghastly as these islands of immigration were, they were still islands that supplied hope, the possibility of better lives; they were still islands of dreams. There were other islands of exile that supplied no such dreams; indeed they were the end of all dreams.

Napoleon, for example, was a man of islands, born in Corsica, an island that he soon deserted. He was to know others: Elba, where he was sent first, and finally St. Helena, where death came. They say that Napoleon had written as a boy when asked for an essay, "*Sainte Hélène, petite île*"—and that was all. Unusually meaningful words. He had been born on an island; on which one would he die? There is certainly evidence that he had little affection for any island, that indeed even the world itself was not big enough for him. While he was at Elba he dreamed and those who were left behind in Paris and throughout all of France, waiting for his return at the time of the violets, dreamed with him. But there was no such hope or dreaming in St. Helena.

Napoleon's biographer, the Marquis de Las Casas—his Boswell—tells us of some of these days:

ISLANDS

November 4, 1816. Today the Emperor would not receive any one during the whole morning. He sent for me at the hour he had appointed for taking the bath, during which, and for some time after, he conversed on the knowledge of the ancients, and the historians by which it has been transmitted to modern times. His reflections on the subject, all led to the conclusion, that the world was yet in its infancy. . . . We then took a view of the structure of the globe. . . . I calculated that Europe contained 170,000,000 of inhabitants. The Emperor remarked that he himself had governed 80,000,000; and I added that, after the alliance with Prussia, he had marched at the head of more than 100,000,000.

Afterwards, when speaking of the wonders of his life, and the vicissitudes of his fortune, the Emperor remarked that he ought to have died at Moscow; because, at that time, his military glory had experienced no reverse, and his political career was unexampled in the history of the world. He then drew one of those rapid and animating pictures which he sketched off with so much facility. Observing that the countenance of one of the individuals who happened to be present, was not exactly expressive of approbation, he said, "This is not your opinion? You do not think I ought to have closed my career at Moscow?"

"No Sire," was the reply; "for in that case, history would have been deprived of the return from Elba; of the most generous and most heroic act that ever man performed; of the grandest and most sublime event that the world ever witnessed."

"Well," returned the Emperor, "there may be some truth in that; but what say you to Waterloo? Ought I not to have perished there?"

"Sire," said the person whom he addressed, "if I have obtained pardon for Moscow, I do not see why I should not ask it for Waterloo also. The future is beyond the will and the power of man; it is in the hands of God alone."

176

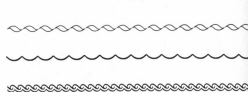

10

Among Some Sporting Islands

HANGING from hooks in the stalls of the Jefferson Market throughout the nineteenth century were the wild fowl and birds that would supply the gourmet tables of New York City. The majority came from the then great sporting island, Long Island, and its sister island, Fire Island. Scattered particularly along the South Shore of Long Island and on the bay side of Fire Island there were old hunter and baymen camps as well as more elaborate clubs for the well-known hunters of the nineteenth century who came from all over the country to shoot and fish.

Thomas Devoe, the director of all the markets of New York City, was an expert on wild fowl and game of all kinds and was well known to the old baymen of Long Island as he personally purchased most of the game that came to the city. He was, however, alert to what could be by the turn of the century the terrible depopulation of birdlife and within the nineteenth century began to warn:

In naming the numerous species of game and other birds, I am anxious merely to show those which it is proper to take, kill, or destroy, as well as those which directly or indirectly affect our tables or are found in our public markets. I doing so, I do not wish to encourage the destruction of a single life that would be more useful to the economy of nature than its dead body for the table. In fact, I would go so far as to wish the passage of a United States general law that would especially protect all birds smaller than the quail, except a few shore-birds, or those which are considered and known to be injurious. . . .

The summer duck or wood duck.

178

Wilson's snipe or English snipe.

American osprey.

His old manuscripts carefully tabulate some of the birds that came to his attention, some so remarkable that they needed special annotations. There was, for example, the wild swan, or whistling swan, that in those days was shot less than a year old for the table. Those were not the days in which such birds would be admired from afar, but they were summarily shot when seen.

In 1846 the *Brooklyn Star* reported that "a flock of white swans made their appearance in Hempstead Bay and one of them was shot weighing seventeen pounds." Some years later on February 25, 1866, a member of the Sammis family of Seaside House, one of the famous resort spots of the island, found two such swans and bagged them in Rockaway, Long Island. They were worthy of memory because their wings measured eight feet extended and they weighed some twenty-five pounds.

The trumpeter swan came not so much from Long Island as from the small islands in the Chesapeake Bay. They were considered, however, dry and tough. The wild goose, or Canadian goose, was plentiful. Again, they were eaten only when young, but many had been tamed, for example, around Spuyten Duyvil Creek and it was the domesticated version that was coming to the table.

The brant goose was considered a fine bird, but a little "sedgy" because of the marshlands it fed on. The canvasback duck was highly considered, particularly in the islands from around the Chesapeake, Potomac, Delaware, and Susquehanna where it fed on the wild celery and therefore had an unusually superb flavor. It was the canvasback that was sent commonly to Europe on those "swift steamers" of the time.

179

The old gunners were not always knowledgeable about birds. Ducks would be called by one name in one area and quite differently in another. The old baymen in particular and the rivermen of the Hudson in the nineteenth century often confused broadbills and redheads. They would bring them to the market, where they would always receive the same price, twenty-five cents. As time went on, the canvasback became the most popular and that bird soon became worth two or three dollars. The mallard, the black duck or dusty duck, the wood duck or summer duck, and the baldpate or American widgeon were all "highly esteemed." The broadbill or bluebill or scaup duck, the creek broadbill, and the bastard broadbill were considered "good eating only when fat." On the East Coast the gray duck or Welsh drake was rarer, but spoonbills were still common and the weaser, an old bay name, would occasionally turn up in the market of Abraham Snediker, a specialist in game birds. The old wife or old squaw, another name for the long-tailed duck, had to be cooked properly but always sold at a low price. The skunk duck or sandshoal duck, very fishy, was less popular. The white-winged coot was "an indifferent bird" for the table, as was the surf duck. The hell-divers and the American scoters were common, but considered poor eating.

Occasionally a loon would be seen, but they were purchased more by those who collected wild birds for their stuffed-bird collection than by any gourmet. In the nineteenth century strange bits of bird lore were always appearing in the newspapers under the heading "Curious Fact." One loon came to grief with the steamer *Oregon,* en route from Stonington, Connecticut, to Huntington, Long Island:

While off Huntington, one of her injection pipes suddenly became stopped, making the engine falter considerably. Mr. Lockwood, the assistant-engineer, observed the circumstance immediately, opened the spare injection-pipe, and directed the attention of the chief-engineer, Mr. Vanderbilt, to the stoppage of the other one. Mr. Vanderbilt supposed the stoppage was occasioned by the boat going over some seaweed, and would soon get clear again. It not doing so, however, he examined the cause of the stoppage when the boat arrived at New York. After taking off the injection valve and a portion of the pipe, he found in it, tight up against the guard of the valve, a large (loon) duck, weighing seven pounds, which had been drawn into it by the force of the vacuum created by the engine. Mr. Vanderbilt thinks that the duck must have dived when the boat approached it, as, when it was found, its head was downward, with its back towards the bow of the vessel.*

180

The bird was stuffed and presented to Barnum.

In 1821 a pair of grouse sold for five dollars. A few years later they were selling for eight and ten dollars a pair, but by 1861 they had dropped to a low price of about fifty cents a pair, the island bird market having been glutted by birds from the Midwest which were being shipped for the winter and early spring months.

The English snipe, Wilson's snipe or common snipe was excellent, considered only second to the woodcock as a game bird, and it would now and then occasion a small adventure. For example, J. T. Brownwere hunted for snipe in a swamp near Bushwick Ferry, Long Island, in 1851 and ran across rather strange fauna for the area:

"While in the act of leveling his piece at a flock of snipes he discovered an alligator within a few yards of the spot where he stood, making towards him, when he instantly lodged the contents of the piece in its throat and killed it."

The meadow snipe or pectoral sandpiper known by the baymen in Long Island as the shortneck, and by the islanders around Barnegat Bay, New Jersey, as the fat bird, was "much sought after." A bird often shot

* *Commercial Advertiser,* April 30, 1847.

and rarely eaten was the long-billed curlew, one of which was shot in 1838 by Daniel Fordham of Southampton, Long Island. The seaside finch was familiar to the old baymen, but as it was a fishy bird it was not liked too well in the market. The black gull and the black tern were not eaten, but collected.

Perhaps the most fascinating of the stories of rare birds are the stories of the bird that has become nonexistent, the passenger pigeon. One of the old documents of Devoe's discusses that bird in detail, and it is not difficult to see how we lost forever one of our common species:

These numerous birds are found in our markets, both alive and dead, very plenty, and generally cheap in the latter part of September and October; they are also found in less numbers through the winter months.

Great numbers are taken alive with nets, cooped up for several weeks, and fed with grain until fat, then brought to our markets as the prices advance; while those that are brought dead have been shot from off the "spar," and sent here at the time of their "flying," which takes place generally in the month of March, when they are going north, and then again in the fall, about the 15th to the 25th September, when they leave for the Southern climate. Large numbers of the old birds and squabs are sent here from the "West," where they are killed or taken alive at their "roosts." The wild squabs, when fat and fresh, are very delicate eating; the cooped bird is also good, the flesh being rather dry; but a poor wild-pigeon is very indifferent eating, even if well and properly cooked. They are found best in the months of September and October.

I have often enjoyed the sport of taking wild pigeon, both with the "net" and shooting them from the "spar. . . ." I have known fifty killed at one shot. I chanced to kill seventeen with one barrel, and thought I was doing considerable in that line.

The largest number taken in a net, I ever heard or read of, was noticed in one of the Detroit papers—the *Owosso American*—in the year 1858, which says: "Mr. Merritt Richardson, living a couple of miles south of this village, on Wednesday last, caught, at one haul of his net, six hundred and forty-eight wild-pigeons."

In the seasons of "great flights," thousands are brought to our markets; and, in "olden time," above one hundred years ago, they were sold very cheap. I find, from the *New York Mercury*—"One day last week, upwards of seventy-five thousand pigeons were brought to the market, insomuch that fifty were sold for one shilling."

The *Boston Weekly Post-boy*, May 2, 1771, says: "The great numbers of pigeons that have been brought to our market within the fortnight past has greatly reduced the prices of all kinds of provisions. It is said that nearly fifty thousand were sold in one day."

181

ISLANDS

Even the small birds, the singing birds, were shot and offered in the market, larks, for example, frequently being sent from Long Island, as well as English robins and finches. Eagles, both the gray and the bald-headed species, were often killed on Long Island.

It was an age, of course when the bird was still a table bird, when everyone would eat what he could hunt. It was related, for example, that Prince Achille Murat, who lived in Florida in the midnineteenth century, said that there was nothing that "swims the water, flies the air, crawls or walks the earth, that I have not served upon my table." Consequently he ate alligator steaks, frog shins, boiled owls, roasted crows, indeed everything except the one animal that confounded him. It was the buzzard. He is quoted as saying: "I tried him fried, I tried him roasted, I tried him stewed and I made soup of him, but the buzzard is no good. I have no prejudice against him, but I cook him every way and I do not like him."

Almost the same areas that were famous for hunting were also famous for fishing. Fish poured into the markets of New York.

Some fish were great game fish, the salmon, for example, from the islands around the Hudson River where Hudson's mate in 1609 on the *Half-Moon* had caught the first one of the voyage: "We had . . . a great store of salmon in the river."

182

In 1771 a law was passed, however, against taking salmon in the river, but they continued to be taken and the old rhymes for both the shad and the salmon appeared in popular culture. In a New York paper there was a headline "In January by a shad the mighty city was made glad," followed by "A salmon eel in January made Manhattan quite as merry."

It was around the islands of the Connecticut River, however, that the best shad appeared, superior in size with their square-shaped back, a fatter and more beautiful fish altogether. In Connecticut they were smoked and cured along the shore.

Fish came to the market from farther away too—from the Thousand Islands, for example—where the glorious pickerel and the muskelunge, the kingfish of the lakes, were caught—and from the islands of the Great Lakes. Zeno C. Scott eulogized trolling among the Thousand Islands in 1875:

> Here is the angler's paradise,
>   A dreaming, Eden-like retreat,
> With balmy perfume in the air,
>   And wild-flowers springing at the feet.

All the charms which angling for pickerel confer are sublimated and condensed into trolling among the Thousand Islands. The pickerel of the

Fly-fishing for trout.

Fishing for striped bass.

183

The fishing fleet of Saint Pierre, off Newfoundland.

ISLANDS

"A Maine Inlet," painting by the American artist Fitz Hugh Lane. (Courtesy Museum of Fine Arts, Boston, Karolik Collection)

thousand lucent streams and rapids, shaded by as many floral islands, are much better flavored than are those which dream out an indolent existence while watching for frogs among the lily-pads, or darting, until they wear themselves thin, after the minnows of ponds and rivers.

The Thousand Islands extend from Cape Vincent to a few miles below Alexandria Bay, or about thirty miles, and the average width of river is about five miles. Imagination may better picture than I can describe the hundred and fifty miles of trolling and casting the fly on streams dividing picturesque islands, or islets covered with greensward and enlivened by wild-flowers. Some of these isles are decked with large clumps of copse and grove, and others with stately trees which reach sublimely heavenward. This charming scene is enlivened by the wood-duck and other birds of gay plumage or melodious song. I venture the statement that it is unequaled any where on earth for its beauty, variety, and life of scenery. Neither the water streets of Venice with their gondolas, nor the Bois de Boulogne with its ornamental drives and picturesque lakes and fountains, are at all comparable with the Thousand Islands. . . .

On visiting the Thousand Islands for a few days' recreation, my advice is to go in pairs. A gentleman companion will answer, but a lady is better. Clayton, which is a town nearly midway of the islands, on the south side of the river, is said to be the most convenient point to select for trolling; for, in addition to the best grounds being near there, its central location enables anglers to make a trip up or down the river to the extremity of the islands and to return the same day. The hotels along the Thousand Islands are generally comfortable, and the landlords reliable. Make known your wants to the proprietor, and he will engage a man and boat for you. All the trolling-boats are superior in model for speed and comfort. The boatman furnishes rods, lines, baits, and rows his own boat. I prefer to use my own tackle, even to spoons and feathered squids. Each row-boat is furnished with two cushioned arm-chairs, in which yourself and lady are seated near the stern and facing it. The bottom of the boat is carpeted, and crimson is the favorite color. The fishing-rods are so set, by appliances in the boat and on the taff-rail, that the troll follows outside of the track, as the rods are held at right angles with the boat, like outriggers. The line is from fifteen to twenty yards long, and the troller lets it run from the reel as the gaffer rows along. The trollers soon become so enraptured with the varied beauties of the shifting scenes that they lose the consciousness of being on a fishing excursion until the oarsman calls loudly, "Bite on the lady!" which sufficiently disenchants them for the lady to reel in a pickerel or black bass, or perchance a maskinonge; when "Bite on the gentleman!" is heard, and he reels in a fish to the gaff or landing-net. . . .

The area around Newport, particularly around West Island, was famous for its bluefish, as was Cuttyhunk. The famous West Island Bassing Club had been formed early in the nineteenth century just for the angling of striped bass. A famous old gaffer was called Mosier and it was he who could call up bass the way a farmer brings his chicken to feed. Sometimes Mosier would have difficulty: "I've thrown in the chum of six fish and them scups and cachokset comes up and takes it just for all the world as if they was game! And I hain't seen nothin' of no bass yet."

One place for casting was Snecker's Gap; the fish, as the old guides used to explain, were "reether sassy there on the young flood." But the old guides had other directions: Keep a lookout for the ninth wave, particularly when you are sitting on the rocks. "Don't get down toward a gulch but watch where the waves throw most water when they break for it allers depends on the course of the wind." When the fish were sighted Mosier had new directions:

I see his mate a keeping alongside of him all the time; she's 'bout as big as the hooked one. I mean to gaff that one first. How like tarnation the feller fights, an tries to whip out the hook with his tail; that shows he's gitting tired. When they curl themselves up on the top of the water so that you can't budge 'em, you had better be careful not to hold so hard as to let 'em break the line with their tail, nor cut it off with their back fin; nor so loose as to let him git slack line to unhook, or knock the hook out of his jaw with his tail. There! see him straighten out! He has made his last fight,

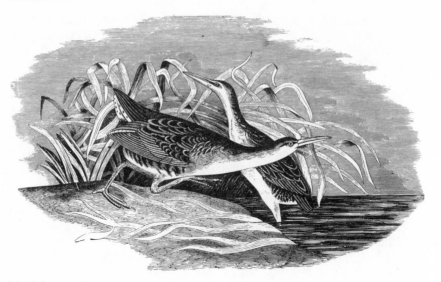

Marsh hen.

186 and got whipped! His mate has gone. 'Twas no use for her to stay an try to help him any longer, for she knows he's dead. Now, with the heave and haul of the tide, there is more danger of breaking the line an' losing him than if he was alive; but here he comes, an here goes the gaff—a forty-pounder at least!

After a successful landing of two bass, members of the club would regularly go to the breakfast table where tea and coffee were served along with broiled bluefish, striped bass and scuppaug, broiled chicken and beefsteak and the famous old bassing stories of the West Island Club.

Then the tour would go over to Pugne Island, owned in the nineteenth century by John Anderson, a millionaire who gave up all for his island home and the finest fishing, he said, in the land. Finally, sailing six miles distant, one reached Pasque Island, described by Scott:

Here we found a club-house with appointments calculated to render not only the members of the club and their families comfortable, but all such guests as members of the association think proper to extend invitations to. The island includes more than a thousand acres, which the club has divided into two farms, erected commodious buildings, including clubhouse, ice-house, stabling, etc. The club has also vegetable and flower gardens, sail-boats and row-boats, and the river, which sets back a mile into the island, is stocked with a hundred thousand menhaden as bait for the use of the club. . . .

What scene can be more refreshing and exalting than an expansive view of the mighty waves, dotted here and there with such beautiful islands

Crying Bird

Frigate Bird

187

Great Auk

Penguin

188

Wild pigeon.

Clam digging at Coney Island where the great boardwalk now stands. (1881)

Lobsterman checking his pots near Porcupine Island, Acadia National Park. (National Park Service photograph by Cecil W. Stoughton)

as those in the Vineyard Sound? The Elizabeth Islands offer the condiments of existence to season the dry hurry-scurry and commonplaceism of the business world on the main lands of America; and they will, before many years, be numbered with the watering-places of the world par excellence.

The trout fishing on Long Island was superb. At one time the greatest angling was at old Obadiah Snedecker's, where even Daniel Webster made sure that he would be for opening day. But the Snedecker Preserve, as it was called, then passed into the hands of another sporting club called The Southside Club.

But one of the finest trout preserves in the United States was on Lake Massapequa at South Oyster Bay on Long Island, on the estate owned by William Floyd Jones. It covered some eighty acres and was fed by a spring brook seven miles in length. Jones maintained it not so much as a private club, but strictly for the use of his invited guests. Both a fish preservist and a fish culturist, he was also a first-class sportsman whose horses were known as much as his streams. That great fisherman, Scott, used to say about trouting on Long Island:

Trouting on Long Island is the most artistic angling that I have ever seen practiced, either in Europe or America. The trout there appear to have learned to detect many of the angler's artifices. Fly-fishing is there practiced near the estuaries of streams, where they are influenced by the tides, so that in flood tide the fisher begins below and casts along as the tide makes as far up the stream as the trout feed; and when the tide turns, the angler fishes along down with the tide and the feeding fish. There being little protection to veil the angler from the tenants of the stream, it is necessary that he keep far back from the bank which necessitates long casts and frequently the first intimation which the angler receives of a bite is the gushing and slapping rise of the fish, and the tremulously nervous resistance at the end of his line; then approaches the play and the contest when light—but finely-constructed—tackle tells. Deftly and gingerly are the words, for Long Island trout are not to be trifled with. The rod should be permitted to do its duty, and the angler be neither impatient nor excited. Anglers who have never visited Long Island are comparatively innocent of the real zest of trouting; for without being annoyed with stinging and biting flies, the trout are as large and as free from rust or the effects of discolored waters as are those of the estuaries on the coast of Maine or along the Gulf of St. Lawrence. On the island they run from a quarter to three pounds in weight, sometimes more, and are in the highest state of succulent adiposity. The climate is charming, surroundings most inviting, hotels where good cheer greets the sportsman throughout the year. I love Long Island and venerate its trout streams.

190

Such lyricism about Long Island has, alas, disappeared along with the trout—and a good deal of the shellfish. Shellfish once abounded in the waters of our islands with a glorious abandon.

In the history of our country, the oyster was the most inviting of the foods in the bays. It was the beginning of an oyster appreciation that reached extraordinary heights in seventeenth, eighteenth, and nineteenth century America.

Some of the first settlers in the United States were suspicious of the oyster. Perhaps the mere fact that he was so lazy and so easy to obtain made him less appreciated. In any case, the early Dutch islanders of New York and the western end of Long Island complained bitterly that they had nothing to eat but oysters. They were the exception, however, because the Dutch had a long and distinguished gastronomical joy of the oyster. Oysters were worthy of great painters and great paintings, and in those Flemish paintings you often see oysters so beautifully painted that they whet the appetite.

The original Dutch in this country ate oysters in a variety of ways. They had delicious pasties, small pies in which the oysters were stewed with parsley, pepper, walnuts, mace butter, and lemon juice. However, one of their favorite ways was the world's favorite, the oyster eaten out of the hand.

The oyster comes to maturity in bed. That undoubtedly accounts ·for easy sensuality. Many foods, of course, could be accused of the same life of indolent existence. Not only does he grow in bed; he remains in bed even as you gaze down at him on an oyster plate. He is a spectacular thing of beauty. Pleasurable to the tongue, he's exciting and inviting to the eye and a gentle invitation to bed.

For the generation, however, that first stepped on the Eastern shore of the United States, the amorous lore of the oyster was far less important than the fact that it was food. The oyster gave hope. When there was no other food, the oyster was there to be taken from the bays.

The oysters of the New World greeted the first settlers with abundance and glory. They covered our country with geographical aplomb. American oysters—*ostre avirginica*—seemed to be found everywhere. They were enormous in the harbors of New York City. They surrounded Governors Island, they named Oyster Bay, they inscribed their importance upon the banks of Long Island.

In Great South Bay of Long Island there are the Bluepoints and the Fire Island Salts. On the East End there are the Gardiner's Bays and the Gardiner's Island Salts. There are the Peconic oysters, the Sea-pures from the deepest waters with high salt flavors. Some Long Island oysters have heavy shells with sweeter flavors and harder meat, some have creamy interiors and pearly shells, but all are magnificent.

191

The lore of the oyster is a regional lore, however. We know old oystermen on Long Island who from the juice alone can identify every oyster blindfolded. Some of the lore even gets into conversation. A "Sea-pure fellow" is a salty companion; a companion who's as "hardshelled as an Oyster Bay" is likely to be insensitive.

Rhode Island has its lore and oysters: the superb Silver Leaf, looking almost like a scallop; the Narragansett Bays, surprisingly sweet for the brackish water in which they make their beds, and the saltier Sea-to-home. The fat and sassy Nyatt Points were nurtured on sunlight in the once clear waters.

Chesapeake Bay and its islands are world-famous for their oysters. They have some of the romance of the handsome old pungies, the oyster boats that used to be touched with Caribbean abandon, a touch of pungy pink here with bronze-green bends and white rails. The old pungies are gone, but the last all-sail fleet (mostly skipjacks) of commercial craft in the United States is there in Chesapeake Bay. Dredged only by sailing ships, the oysters that have emerged are as fragrant and masterful as the old recipe for pickled oysters that you can find around the Chesapeake:

Newfoundland fishing village, Parish Cove. (Courtesy Canadian National Railways)

192

Strain the liquid to which you add (as did the Dutch) a pinch of mace, a pinch of salt, a pinch of pepper. Allow liquid to boil and insert oysters until edges curl with pleasure.

Once again separate oysters from their liquor and cool both. To a gallon of oysters [yes, that's how you once got them in the Chesapeake] you add a cup of vinegar and a cup of white wine plus the original liquor. An overnight bath in this fragrance, and the oysters are served.

(All inland areas of the United States knew pickled oysters. Even areas only a short distance from the sea had to pickle them. Oliver Wendell Holmes tells us of a dinner party where one statement "Shells—not pickled" brought everyone racing to the oysters.)

The Choptanks, oysters of the Choptank River, which enters Chesapeake Bay, have a history of sweet waters (so much of our sweet water is disappearing) and sweet mystery. The mammoth oysters, the Chincoteagues, are world-famous; together with Bluepoints, they were favorites at Prunier's in Paris.

Oysters don't like the cold. Some of the sweetest, if scarcest, are found in the South. Coon oysters used to be a joy in the Carolina islands. These almost tree-growing oysters account for the ancient stories of trees

Weighing in at 74½ pounds, Allison tuna like this one are common to Bermuda's fishing grounds. The colony's fishing is ranked among the tops in the world. (Courtesy Bermuda News Bureau)

193

that bore barnacles and produced the fabulous animal of antiquity, the barnacle goose. Even such a well-intentioned authority as Gerarde, as late as 1597, declared in his *Historie of Plants* that such trees delivered forth shells that produced a goose.

Oysters grow on the mangrove islands of Florida. When we were children, we wintered near such oyster mounds. At high tide we swam over them, scratching our legs against the rough surface as vicious as coral. At low tide we gathered them, greedier than any raccoon.

We preferred them in one of Johnnie MacDonald's oyster roasts. He was the best Cracker oyster roaster around, and acquired his knowledge of Florida waters from an old Seminole. Johnnie allowed a dozen apiece.

First he hosed off the oysters. Then he placed one-inch wire mesh taut across an open fire, anchoring it carefully on all sides, provided that it was loose enough to be shaken occasionally with tongs. The oysters were placed discreetly on the wire, not piled up, even if the children were ravenous. When the oysters sprang open, we sprang at them. On rainy days we cooked them inside. Johnnie didn't "hold fit" to cooking oysters inside. "The only house an oyster wants is a shell," he maintained, but when pressed he would serve Web Givens's Florida oysters.

WEB'S FLORIDA OYSTERS

36 shell oysters	¼ tsp. paprika
½ cup bread crumbs	6 slices bacon
½ tsp. salt	enough freshly ground pepper
juice of 2 limes	to see it

Shuck and drain oysters. (Johnnie always shucked oysters by placing them on a flat surface. He used the back of his pickup truck. The flat shell should be up. Break off the "bill" or the thin end with a hammer. Legend says that the Seminoles did this with their teeth. Next cut the abductor muscle. This is found close to the flat upper shell. Remove shell. Cut the lower end of muscle in lower shell, but leave oyster loose in lower shell.)

After oysters are drained, mix bread crumbs and seasoning, and sprinkle over oysters. Cover oysters with pieces of bacon. Place in preheated oven, 3" from broiler heat. Broil for five minutes until edges curl. Serves six.

The barrier beaches on the islands of New Jersey were particularly famous for oyster roasts on the beach. There, blessed with sand and ceremony, the oysters broiled away over driftwood fires while children waited and danced in the shadows.

194 The snail, the crab, the clam, and the lobster—or, in other countries, the crayfish, the mussel, and the winkle—are rich in legend, the history of eating customs, and even religious significance. Just as the oysters were considered a cure-all from the time of Seneca, the crab even today is a cure for earache in the Bahamas. You simply pour the water from the claw of a crab into the ear, and naturally the ear gets better. Conjure men in South Carolina always used crab claws to make themselves invulnerable. And if the first men, so some of the legends of the Pacific said, had eaten crabs instead of bananas, all of us would be immortal.

The old sporting clubs have left us with a superb collection of old recipes—none better than Daniel Webster's chowder from Long Island:

4 table-spoonfuls of onions, fried with pork.
1 quart of boiled potatoes, well mashed.
1½ lbs. sea-biscuit, broken.
1 tea-spoonful of thyme, mixed with one of summer savory.
½ bottle mushroom catsup.
1 bottle of port or claret.
½ nutmeg, grated.
A few cloves, mace, and allspice.
6 lbs. fish, sea bass or cod, cut in slices.
25 oysters, a little black pepper, and a few slices of lemon.

The whole put in a pot and covered with an inch of water, boiled for an hour and gently stirred.

MAJOR HENSHAW'S CHOWDER

Cut up a pound and a half or two pounds of old salt pork into small pieces, and put it in a pot that has a close cover. Put in four table-spoonfuls of sliced onions when the pork is nearly tried out, and when the pork is entirely tried out remove the pieces with a skimmer or large spoon.

Then take six pounds of sea or striped bass, cod, or any other firm fish, and cut it in slices; a pound and a half of broken biscuit; twenty-five large or fifty small oysters (these may be omitted if out of season); one quart of boiled potatoes well mashed; half a dozen large, or eight or ten small tomatoes sliced (or half a bottle tomato catsup instead); one bottle port or claret, or other wine (the two former are best); half a nutmeg grated, a tea-spoonful each of fine summer savory and thyme, and a few cloves, mace, allspice, black pepper, and slices of lemon. Put the first five articles in the pot in layers, and alternately, in the order above stated; sprinkle over each layer a portion of each of the other ingredients, then put in water enough to cover all. Cover close, and let it simmer, and stir occasionally till done. It should not boil, but·simmer slowly, and the cover should be taken off as seldom as possible; on this the flavor depends. When the fish on top is done, serve up the chowder.

195

BASSING CLUB CLAM CHOWDER

Butter the bottom and sides of a deep tin or earthen dish; strew the bottom thickly with bread crumbs or rolled cracker (soaked); sprinkle over it pepper and pieces of butter the size of a hickory-nut, and parsley chopped fine; then put in a double layer of clams. Sprinkle also over them pepper and pieces of butter, then another layer of soaked crumbs or cracker, and again a double layer of clams, pepper, butter, and so on, the last layer being of crumbs; add, finally, a cup of milk, or, in lieu of it, water. Put a plate over the top, with coals above and below, or bake in an oven three quarters of an hour. If too dry, before it is done add enough milk or water to moisten it.

Fifty clams, half a pound of soda biscuit or bread crumbs, and a quarter of a pound of butter, is the quantity necessary for this receipt.

TO FRY SALT PORK NICELY
(SOUTHSIDE CLUB)

Cut it in thin slices, and put it in a fry-pan covered with hot water; let it boil up once, and then pour it off; shake a little pepper over it; let it fry on both sides in its own fat, then ·take out the pieces and add to the

196

Isle Royale National Park on Lake Superior. A commercial fisherman, longtime resident of Isle Royale, tends to his nets and drying frames. (National Park Service photograph by Jack E. Boucher)

gravy a large teaspoon of flour; stir it till smooth and free of lumps; then add a cup of milk; stir over the fire a few minutes; shake more pepper over it; then pour it over the pork, and serve; or thin-sliced boiled potatoes, or fried or boiled cold parsnips, may be fried in the gravy when the pork is taken out.

CLAM OR OYSTER FRITTERS

Open and dry them with a towel; mix two well-beaten eggs, somewhat less than half a pint of thin liquor and half a pint of milk (or the same quantity of liquor in addition if you have no milk), with a pint of flour; beat it thoroughly together till it is free from lumps; then stir in the clams or oysters; cut up some salt pork in small pieces, and try it out in a fry-pan, and remove the pieces of pork. When the fat is boiling hot, put in your clams or oysters with a large spoon, with one or two clams, etc., and batter in each spoonful. Let them brown, and then turn them over; as soon as done, remove them from the pan, and lay them on a gridiron with a dish under it to catch the drippings. There should always be enough fat in the pan to cover, or nearly cover the fritters.

Punches were found on every island in every bay camp and fortunately a few of those old recipes have been preserved. It was after a day of drinking flips on Fire Island that one old bayman saw the only merman to have visited our shores.

SHERRY-COBBLER

Put in a tumbler a table-spoonful and a half of powdered sugar and a slice or two of lemon; then fill it half full of crushed ice; then pour on it a wine-glassful or more of sherry. Pour the whole from tumbler to tumbler till well mixed, and drink through a straw, if you have it.

MULLED CIDER

Take a pint of sweet cider; reserve a tea-cupful of it, and add to the remainder an equal quantity of water. Set it to boil, with a tea-spoonful of whole allspice added to it; then beat three eggs very light, and stir gradually the reserved cup of cider into them; then stir this mixture gradually into the boiling cider and water, and continue stirring till the whole is smooth; sweeten to taste; grate a little nutmeg over it, and serve hot in tumblers.

197

FLIP

Put the quantity of ale, porter, or beer you wish in a tin cup, and add sugar to taste; heat the end of a thick piece of iron red hot, plunge it in the liquor, and stir round till the liquor ceases to bubble, and drink hot. This is the most refreshing and strengthening drink either before or after a hard day's hunt that I know of. A piece of iron of the shape and size of a large soldering-iron is the best.

One old recipe adds: "Although sportsmen and mariners do not seek either the wilds or the waves for the luxuries of the table, yet they set a higher estimate on heaven's bounties than to suppose meat and drink given to sustain life only. They consider them rather as bestowals for strength and enjoyment to man, and as such they are to be used intellectually and in moderation."—God bless!

11

Islands of Lost Animals and Birds

It is a collection of seven islands, graced and gutted with sea cliffs, with shards of rock projecting into the sky like manmade lighthouses. To naturalists all over the world this nest of islands is known as St. Kilda and it was, until destroyed by man, one of the last breeding places of an almost mythological bird, the great auk or garefowl. It was a handsome bird, a full thirty inches high, glistening black, except for a round circle like an eye of Medusa between the beak and the piercing eyes. Indeed it looked, at least to those who had seen both birds, somewhat similar to the penguin, and like the penguin, its walk on the rocks seemed both sure and unstable at the same time. But those shining black wings were a deception; the body was far too large for the wings, and the bird was flightless.

And yet it existed for years; the splayed feet made it a magnificent swimmer. It took to the waves the way a ship might, and like a ship it came into these rocks of St. Kilda for years to breed and then in one terrible year, it came to die.

Off the north end of Boreray is the stack of rock called Stac an Arminn. Here was found and killed the last great British auk.

The islands have always been isolated; the rocks are so treacherous, the locations so difficult to reach that in effect the area had never been mapped by the British Ordinance Survey. It wasn't until 1927 that it was privately surveyed, and although these mountains in the sea have only somewhat more than two thousand acres it took two men five months to survey them.

Even in the seventeenth century they attracted naturalists. Martin Martin of Skye, the tutor to the children of the head of the Clan of MacLeod, the hereditary owner of the islands, wrote of his visit in A Late Voyage to St. Kilda in the middle and the end of the seventeenth century. He was fascinated by the clouds and white mist that crowned the rocks and fascinated, too, by a rare little bird, the wren. There were four unique animals on St. Kilda: the St. Kilda wren, the St. Kilda sheep, the St. Kilda fieldmouse, and now extinct, or probably so, the St. Kilda house mouse; in addition there were innumerable gannets and puffins.

When Martin arrived, there were 180 people living on the treacherous rocks. They lived completely on birds. They climbed over the cliffs as surefooted as most rock island hunters. In the time of Martin they ate the gannet; a little later they ate the fulmar. In the nineteenth century an anonymous visitor described in detail the wild-fowling at St. Kilda:

The solan-goose is not a very wise animal, and is, generally speaking, easily caught. Here is one of the ways by which these birds are taken: they are very fond of herring, and if one of these fish be fastened upon a board, and then set afloat, the bird, only taking note of the fish, sweeps down on it with such eagerness and force as to drive its beak through the herring into the piece of wood, and, if not at once killed by the concussion, is kept a prisoner till his captors can take possession of him.

The various stacks which are accessible to the people of St. Kilda are all inhabited by sea-fowl, the solan-goose or gannet being by far the most numerous. These birds do not live on these islands all the year round, but only use them to breed upon, when they remain two or three months and then depart. . . .

It is not easy to convey to those ignorant of such matters an idea of the enormous wild-fowl population which inhabit the beetling cliffs of St. Kilda and the adjacent stacks of Soa and the Dun, or the further away rocks of Narnin and Leath. The sea-birds of these places will number a million or two. This will be obvious when it is stated that the islanders have been known to capture as many as twenty thousand birds in a few weeks. Even when seen from a great distance, the rocks seem as white with gannets as the snowy peaks of Mont Blanc do with the eternal snow which adorns them; and when a gun is fired to frighten the birds, and myriads leave their nests in great alarm, the young which remain keep the aspect of the rocks unaltered—they look as white as ever.

It is almost impossible for any drawing to delineate effectually the dangers which are undergone by the fowlers who carry away the birds, but [one of the pictures in this chapter] is a faint attempt to show how fowling is carried on at St. Kilda and the neighboring rocks. The "island of the

199

The wood ibis at pelican refuge, Florida, has a wingspread of five and one-half feet. (Bureau of Sport Fisheries and Wildlife photograph by Luther C. Goldman)

200

The attack on the seals. (Painting by Henry Wood Elliott. Courtesy of The Robert Lowie Museum, University of California)

gannets," as the rock of Leath is called, presents in some places a sheer wall of nearly smooth stone, rising for hundreds of feet perpendicularly out of the sea. To this place the men of St. Kilda, to the number of sixteen, will proceed in their boat, and, climbing to the very summit of the beetling cliff, or to some part of it on which they can obtain a tolerably firm foothold, will then, by the aid of strong cords, descend in search of their prey, which having secured with great rapidity and dexterity, they are quickly drawn again to the summit, or the place from which they made their descent. It takes away the breath of those unaccustomed to such sights to see the daring which these men display whilst engaged in their calling. A hundred feet above them is their place of safety, and five hundred feet below the ravening water is ready to receive them should the rope by which they are suspended happen to break. Happily accidents are almost unknown. At any considerable distance the men appear almost no bigger than so many large spiders crawling upon the wall of rock. It requires a field-glass to discern the minutiae of their business, and to see how deftly they seize the birds, and how quickly and ruthlessly they kill them. When looked at from a boat, about a quarter of a mile distant from the cliff, the men may be seen waving a small fluttering white object, just like a handkerchief. It is, however, a bird they have seized, and in a moment it is dead, and looped on the string which is attached to the main rope. . . . As has been said, accidents rarely occur, even although the ropes are doubly laden, and one man hangs below his fellow, the same cord serving for the use of each of them. Once, when two men had descended from a high cliff, both suspended by the same rope, the man who was uppermost saw, to his horror, that one or two strands of the cord had given way, and that the death of both of them was almost inevitable. There was not a moment to lose; quick as thought the man severed the rope that held his companion just beneath the part where it was fixed to his own body, and in a moment the poor fellow was engulfed in the waters which were ravening far below him. The act was inevitable, and, even by committing it, the man barely saved his own life. Had he not been promptly seized by those above as he was drawn to the top of the cliff, he, too, would have paid the forfeit of his life upon that awful occasion. . . .

201

The people of the island live chiefly on the puffin, as likewise the eggs of the sea-birds which frequent St. Kilda. They have also a flock of two thousand sheep and about thirty or forty cows; they shear the sheep and dress the wool, weaving it into a kind of homely cloth of the Tweed kind, which is of considerable value to them. But their great industry is evoked in the way of obtaining feathers which are plucked from the fowls taken from the rocks, entailing the annual capture of thousands of these winged animals. And thus, toiling on and on, the people live, repeating year after year the homely and uneventful story of their lives:

> With sparing temperance at the needful time
> They drain the scented spring; or hunger-prest

> Along the Atlantic rock undaunted climb,
>   And of its eggs despoil the solan's nest,
> Thus blest in primal innocence they live,
>   Sufficed and happy with that frugal fare
> Which faithful toil and hourly danger give.

Eventually the birds began to give out. Such naturalists as W. H. Hudson warned against the fate of the St. Kilda wren, its eggs so valuable to collectors that the islanders picked them up and sold them indiscriminately.

Hungry, defeated by the land, the people began to emigrate. The 180 Gaelic-speaking people moved on; thirty-five left for Australia, others left for the mainland and on August 28, 1930, the last St. Kildans left the island.

But they left behind them a terrible story, the memory of two natives—one MacKinnon, the other MacQueeen—who set out one warm July day in 1840 with one of those magical mists curling around their eyes and fogging their heads. As they climbed the Stac an Arminn they clutched each other. There standing on the rock was a strange sight. Surely it was a witch. Terrified, depleted with hunger, alien to the world they live in, where birds were only a food supply, they had seen no giant bird like this in some time. In the shadows of the mist it must have seemed a witch, so they took up their clubs, beat it unmercifully and killed the last British auk. But their story was simply a footnote to what would come later. Four years later on another island off Iceland, the great auk's story came to a terrible conclusion. A three-masted schooner anchored in the water and released the last predator of the auk—the hunters. The last auk were nesting on the island. The sophisticated hunters saw a bird that they knew was not a witch; they knew it was the garefowl, the great auk, and they clubbed to death the last auk known to man.

Now a hundred years gone, the great auk once traveled our islands in the North American waters, seen as far south as North Carolina, known on Cape Fear, Cape Lookout, Ocracoke Island, around Hatteras, Pamlico Sound, Currituck Sound, Delaware Bay, Block Island, Martha's Vineyard, Nantucket, Race Point, Cape Ann, Portland, Maine, and in the Penobscot and Casco Bays, at Damariscove Island, at Bar Harbor, at Cape Sable, at Cape Breton Island, at Sable Island, at Conception Bay, at Trinity Bay, and in Bonavista. They moved from their summer mating grounds in the Carolinas northward to Newfoundland, then to Belle Isle, to St. Michael's Bay, to Huntington Island, following and occasionally crossing the Labrador Current, then on to Cape Farewell in Greenland and across to Iceland: a three-thousand-mile migration in which they

202

The ring-billed gull, common on East Coast on lakes, bays, harbors, and frequently in plowed fields. (Bureau of Sport Fisheries and Wildlife photograph by Luther C. Goldman)

203

Murres on St. Paul Island, Alaska. (Bureau of Sport Fisheries and Wildlife photograph by V. B. Scheffer)

weathered each year all the vicissitudes of migrating birds, only to die out completely in Iceland.

There is no monument to the last of the birds on that island, Eldey Island, or Fire Island as the natives called it, off Iceland, but there is a kind of monument to the auk and all other birds and animals that have been decimated in recent years, on one of the islands of St. Kilda, where the last British auk was killed.

Now the island is owned by the Marquis of Bute and it is a true naturalist's island, a sanctuary. There you can take no rock or plant, kill no animal or bird; there you can only see and observe. And there are new things to see, now that man has left the island altogether. It is now a laboratory. What happens to an island when man leaves it, having decimated its birds? Will they come again? What new pairs can be discovered? Naturalists have tabulated for us the changes; the otter, the duck, and the gannet are there in quantity, the golden plover is breeding, but still there is much that is not known about the birdlife. Even today one cannot be sure where the petrel and the shearwater nest in the rock. It has been the scene of many expeditions from Oxford and Cambridge. It has been visited by students and by that great naturalist Frank Darling. It has atoned in some way for the death of the auk.

204

So the auk joined the dodo, that remarkable bird which once existed in abundance in the islands of Mauritius, Bourbon and Rodriguez, graphically described by Sir Thomas Herbert, who saw the bird in 1634:

The Dodo comes firft to our defcription. Here and in Dygarrois (and nowhere elfe that I c^d ever fee or heare of) is generated the Dodo. (A Portuguize name it is, and has reference to her fimplenes) a bird which for fhape and rarenefs might be call'd a Phoenix (wer't in Arabia); her body is round and extreame fat, her flow pace begets that corpulencie; few of them weigh leffe than fifty pound: better to the eye than the ftomack. . . . Let's take her picture: her vifage darts forth melancholy, as fenfible of nature's injurie in framing fo great and maffive a body to be directed by fuch fmall and complementall wings, as are unable to hoifte her from the ground, ferving only to prove her a bird; which otherwife might be doubted of. . . . Her eyes be round and fmall, and bright as diamonds; her cloathing is of fineft downe, fuch as ye fee in goflins; her trayne is (like a China beard) of three or foure fhort feythers; her legs thick, and black, and ftrong; her tallons or pounces fharp; her ftomack fiery hot, fo as ftones and yron are eafilie digefted in it.

The nineteenth-century naturalist Phillip Gosse warned that within ten years the auk, the Notornis, and the Nestor had disappeared. A capture of the Notornis was described to the Zoological Society of London in 1850:

It was in the course of last year, on the occasion of my son's second visit to the south of the middle island, that he had the good fortune to secure the recent Notornis [skin], which I now submit, having previously placed it in the hands of the eminent ornithologist Mr. Gould, to figure and describe. This bird was taken by some sealers who were pursuing their avocations in Dusky Bay. Perceiving the trail of a large and unknown bird on the snow, with which the ground was then covered, they followed the footprints till they obtained a sight of the Notornis, which their dogs instantly pursued, and, after a long chase, caught alive in the gully of a sound behind Resolution Island. It ran with great speed, and on being captured uttered loud screams, and fought and struggled violently. It was kept alive three or four days on board the schooner, and then killed, and the body roasted and eaten by the crew, each partaking of the dainty, which was declared to be delicious. The beak and legs were of a bright red colour.

The death of the Nestor, too, was vividly described:

When Norfolk Island—that tiny spot in the Southern Ocean since so stained with human crime and misery—was first discovered, its tall and teeming forests were tenanted by a remarkable Parrot with a very long and slender hooked beak, which lived upon the honey of flowers. It was named Nestor productus. When Mr. Gould visited Australia in his researches into the ornithology of those antipodeal regions, he found the Nestor Parrot absolutely limited to Philip Island, a tiny satellite of Norfolk Island, whose whole circumference is not more than five miles in extent. The war of extermination had been so successful in the larger island that, with the exception of a few specimens preserved in cages, not one was believed to survive. Since then its last retreat has been harried, and Mr. J. H. Gurney thus writes the dirge of the last of the Nestors:

"I have seen the man who exterminated the Nestor productus from Philip Island, he having shot the last of that species left on the island; he informs me that they rarely made use of their wings, except when closely pressed; their mode of progression was by the upper mandible; and whenever he used to go to the island to shoot, he would invariably find them on the ground, except one, which used to be sentry on one of the lower branches of the Araucaria excelsa, and the instant any person landed, they would run to those trees and haul themselves up by the bill, and, as a matter of course, they would there remain till they were shot, or the intruder had left the island. He likewise informed me that there was a large species of hawk that used to commit great havoc amongst them, but what species it was he could not tell me."

Not only birds, but animals, too, were disappearing. Where had the Stelleria—so-called by Cuvier—gone? It was described by Gosse as a giant marine pachyderm, similar to the manatee, but a good twenty-five feet long and peculiar to the Bering Strait and the Aleutians. The ex-

205

206

Sub-adult sea otter on a rock beach, Amchitka Island, Alaska. (U. S. Fish and Wild-life Service photograph by Robert D. Jones, Jr.)

tensive sealing practices were already under abuse by those familiar with them. An anonymous British sealer of the nineteenth century described his aversion to his trade:

"Yes, sir, I can tell you a great deal about 'sealing,' as we call it. I consider it awful butchery: in fact, it is just a case of the roughest kind of killing. But we did not in my day kill mere baby seals, for you see whales were then more plentiful than they are now, so that we were not at the first so anxious about killing seals; and, therefore, we did not begin the fishing so soon as they do now-a-days, and the young ones had time to grow till they were worth killing. . . . No, sir, seals are not so plentiful as they used to be; it is scarcely to be expected that they should be so, when hundreds of thousands of them are killed every spring. At the time they first began to kill the seals, a few thousand was thought a great number. No ship that I ever sailed in to the sealing came home with more than 6,000 and in my day that number was looked upon as a splendid cargo; but now some of the Dundee ships, as I see from the newspapers, will take double or three times that quantity and look for more. And as for the American vessels, if they could kill 50,000 they would take them. I know

Young snake birds at Avery Island, Louisiana. (Courtesy Standard Oil Co. of N. J.)

of a Newfoundland ship that once left Jan Mayen with as many as 33,000 seals aboard of her. I was in the ship and saw the cargo; you could scarce stir your foot for seal skins and scrapings of blubber. You can easily count for yourself what will be the effect on the natural stock if there are two hundred and fifty ships at the sealing every year, each of them killing 10,000 of the animals on an average. Two millions and a-half, do you say? I am sure it will be all that number, if not more. Well, sir, how can such a rate of killing be carried on? People do say that we are catching the herring faster than they can breed, and I believe there is a good deal of truth in that; but the seal, we must remember, is not such a prolific animal as the herring is. I have seen the roe of a herring counted, and the one I saw was made up of twenty-two thousand eggs; but ye see, sir, a seal never has more than one young one at a time. . . .

"Some of the men say that seals understand what is said to them, and think that they might be taught to speak. A shipmate of mine, that belonged to one of the Orkney Islands, believed that seals sometimes throw off their skins and take the shape of human beings. He told me that his grandfather was once married to one. Oh! you may laugh, but he believed it. He said his grandfather was coming home from fishing one moonlight night, when he saw two beautiful females chasing each other on the sand. He was on the point of calling to them when his foot struck on two fine sealskins, which he had no sooner done than the women gave a loud scream and came running towards him. One seized a skin, and putting it on, was swimming in the water in a moment or two; the other one could not get her skin because it was in the hands of his grandfather. She begged and prayed the fisherman to give her the skin, and let her join her companion in the water, but she begged in vain. The man had fallen in love with her,— a case of love at first sight,—and he besought the beauty to come home to his mother's house and procured her some women's clothing, which he persuaded her to put on. In a short time after that they were married, but they were never happy. She frequently asked her husband for her skin, which he refused to give her, fearing that she would desert him as soon as she received it. A settled melancholy fell upon her, she was often observed in a sad and mournful mood, as if there was something preying upon her mind that was as a burden to her, and which she could not, or would not, reveal even to the man who was so fond of her. But for all that, the union of the fisherman and the "silkie wife," as she was called by the Orcadians,

Sea elephants, Channel Islands. (National Park Service photograph by Lowell Sumner)

Native Australians hunting the emu. (Nineteenth-century print from the authors' collection)

Owen's apteryx, from New Zealand. (Nineteenth-century print from the authors' collection)

was blessed with children. The husband, however, saw that the children did not bring a blessing with them, and he began in time to suspect that all was not as it should be. He noticed that his wife cast longing glances on the waves of the sea; he heard strange voices in his house at untimeous hours, and fell, in consequence, into a state of strange unrest. One day he went away as usual to the fishing, but was unable, in consequence of a storm, to get back as soon as usual. He felt anxious, as his boat came near the shore, to know if all was well at home, and looked wistfully at his cottage. As his boat reached the landing shore, he saw two seals on a shelving rock, and at once felt a presentiment that something was wrong. One of the animals, making a sign, thus addressed him: 'Man, you stole away my wife from me, but I am so glad at having got her again that I will not upbraid you; it was only last night that I found her skin where you had hid it, and once again she is mine. I bear you no malice, but you now look upon her who was your wife for the last time, for I shall take her away with me to my bright home in the green crystal sea, there to dwell for evermore.' After the animal had said this, it leapt from the rock and slowly swam away with its companion, leaving the fisherman standing in mute despair. . . .

"It is generally supposed that the sealers never reach the great body of the seals which are somewhere out of reach, but that only the fringes or rear-guard of the main army are seen and killed. . . . We shoot the old seals and club the young ones: they are struck over the nose, which is a mortal part, with a club which has an iron tip upon it. I never could cotton to it; it is like slaying a lot of innocent lambs, the young seals are so helpless, and they have beautiful and beseeching eyes. Then, as for the mothers, I declare they look upon you in such a way as to make you quite nerveless. I can't conceive of a human mother that could look more piteously at the slaying of her young ones than an old seal. Their eyes water with tears, and they gaze upon you so appealingly that you can't help feeling as if you were committing a crime when you kill them. The slaughter goes on as long as we can continue at it; there are not many of the men so sentimental as I am, but the animals, as a general rule, are so helpless that you cannot help thinking how worthless it is to kill them."

The blood on ice and the blood on rock continued. Otters were slaughtered unmercifully on the islands of California, and each October and November the California gray whales were rounded up so successfully in lagoons—25,000 in one century—that they disappeared almost entirely. They are returning now under protection—and islands that were once islands of slaughter are now islands of refuge.

Egg collectors, for example, were not limited to ignorant Hebrideans. Just thirty miles west of San Francisco, the 'Farallons,' little pointed islets in the sea, had known more than their share of "egg wars." The Gold Rush of California had put enormous demands on the available food resources. Why not wild-bird eggs? In the six years between

1850 and 1856, 4,000,000 eggs, nearly all of the common tern variety, were taken by raiders who fought each other as well as the birds. Now the islands are a National Wildlife Refuge.

On the East Coast of our nation Monomoy Island, off Cape Cod, was not prey to eggers but was, as so many of our islands were, prey to the sporting clubs. One area of Monomoy was simply called "Shooters' Island," the home of the Monomoy Brant Club.

Monomoy is now a link in our national bird refuges along the Atlantic flyway from Canada to Florida and the Gulf. It is a particular haven for migrant land birds forced beyond the mainland by strong off-shore winds. Now you can see—not shoot—over 300 species of waterfowl and shore birds.

Moving southward along our barrier islands, we find innumerable refuges in the National Wildlife Refuge System, many supporting animals and birds threatened with extinction—from the delicate Key deer of Big Pine Key to the less amiable American alligator. The oldest "refuge" in our country was Pelican Island in Florida, established in 1903 for brown pelicans, herons, and egrets to protect them from plume hunters and fishermen. Now many islands along all our coasts and in the Great Lakes, as well as many marsh islands of other states, are indeed islands of protection.

211

12

Islands of Lost Civilizations

"Thou hast fathered," they said to Zeus, "a rash and foolish daughter who delights only in guilty acts. All the other gods who live on Olympus obey thee and each submits to thy will, but she, thou never curbest neither by word nor by deed; she does as she pleases."

This is one of the remarks often made by the gods to Zeus about his favorite child, Athena. She had, of course, been born from his own head, but in the island of Crete, where the clouds still hang heavy over the mountains, they said that in truth she had been hidden in a cloud and that when Zeus struck his head against the cloud Athena emerged. Her birth, say some stories, occurred near Knossos, next to the stream The Triton, but in the old mythologies of Greece where the myths merge into each other so that mountains, clouds, seas, and rivers are lost, Athena has other mythological beginnings. Some say that she was the daughter of Poseidon, god of the sea and of the lake Tritonis. When she was born "Great Olympus was profoundly shaken by the dash and the impetuosity of the bright-eyed goddess. The earth echoed with a terrible sound, the sea trembled and its dark waves rose. . . ."

We can read those words now, and we wonder if perhaps Athena had not been born at the time of an earthquake. Earthquakes certainly have haunted Crete and there are many who say that her lands shook when ancient Atlantis sank into the sea. Some old historians felt that it was the lost memory of Crete's greatness that inspired the Atlantis legend. Certainly of all the islands Crete is an island of the sea.

If one were to take one island of Greece and make it a symbol of all others, it would be a great mistake, but if one were to take one island whose history was enmeshed with all of the mythology and history of the Greek islands and mainland and which was as well the repository of a lost civilization, then Crete must be his choice. This, then, was the birthplace of Athena, but Athena like so many of the gods had come from elsewhere. Was she not, perhaps, some Asiatic goddess who when arriving in Greece was still garbed in the fertility snakes of countries more distant than her own mountainous island? Perhaps in the beginning she was a meteorological goddess—a goddess of lightning and of storm; they called her sometimes the goddess of the brilliant eyes. But she was also a warrior and at the same time a goddess of all peaceful crafts and speculative intelligence. She is patroness of the sculptors, as well as the weavers, the builders, spinners, the architects. She guarded the horses and the oxen. She carried the sword that her father had used in the war against the Tritons. She herself wore it against the giants. She protected all warriors—Hercules, who was loathed by Hera; Perseus, who fought against the Gorgons.

213

Crete has often been the center of attention; it has shared many literary, geological, and archaeological quarrels. For a long while it was thought that this was the very cradle of Greek civilization. The discovery, however, of the Linear B tablets made it more apparent (although there are still detractors) that the mysterious language of Crete and its great culture that had appeared almost full grown, so beautiful, so superb that it is one of the great joys of our world today, were indeed subject to the Greek dynasty on the mainland. But still no one completely knows. The beginnings are shrouded in mystery, shrouded in that mythology that seems to be the framework of Crete as much as the mountain backbone of the island. Speaking of Crete, the archaeologist and scholar Jacquetta Hawkes says:

"An island always has a potent effect on its inhabitants. Its frontiers are immutable, divinely determined rather than due to mere human vicissitudes. Strangers cannot easily cross them unnoticed or unopposed. This sense of being 'sea-protected' the envy of less happier lands gives island people a sharp awareness of their identity and of their difference from everyone else. Even today in the Isle of Wight, a modest patch of land fitting close to the English shore, people from the mainland are called 'overners' and are regarded as inferior aliens. An island home, in short, greatly enhances that belief in belonging to a chosen race that is native in every human breast."

Statues, Easter Island.

Minoan ruins at Knossos on the island of Crete. Notice the novel shape of the columns, which become progressively narrower toward the base. (Courtesy Greek National Tourist Office)

The Strait of Chios, the region of Homer's youth.

215

A view of Mt. Etna from Taormina, Sicily. (Courtesy Italian State Tourist Office)

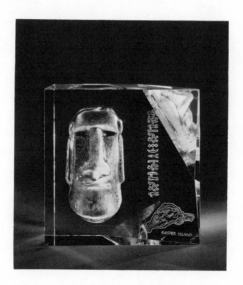

Square crystal monolith revealing the form of one of the mysterious, giant stone heads found on Easter Island. Engravings show a vertical line of still undeciphered hieroglyphics and a map of the island—named by the Dutch navigator Jakob Roggeveen on Easter Day, 1722. (Courtesy Steuben Glass)

216

Crete today is one of the most famous of all archaeological islands. Its stories were old and venerable, and there were those that had read them carefully. One such was Heinrich Schliemann, who was to excavate Troy on the mainland. He knew the story of Theseus and Ariadne, of the strange labyrinth of Minos and the legend of the Minotaur, of Crete's famous pre-Homeric history of being the island of ninety cities. With Germanic literalism Schliemann believed everything he read. The ancient legends had captured his imagination so vividly as a boy that he was determined when he grew up to find out for himself what lay in the ancient Greek mainland and islands. Working as a grocer's assistant, Schliemann accumulated two fortunes, one in Russia and one in the United States. The money would get him to those legendary spots in Greece. He learned both modern and ancient Greek and trained himself in archaeology. Finally, one hundred years ago, he started off for Greece and Turkey. Were the old stories true? Were the *Iliad* and the *Odyssey* just a collection of folktales—the work of one man of superb genius? Did Troy and Knossos exist? Finally as he excavated Troy he stood absolutely enchanted. "I confess," he said, "I could hardly control my emotions when I saw before me the immense Plain of Troy, whose image had hovered before me even in the dreams of my earliest childhood."

Double-bladed crystal ax supported on a golden shaft. The shaft is set in a crystal base curved like the horns of a bull. Engraving on the blade relates the Cretan myth of Ariadne, Theseus, and the Minotaur. (Courtesy Steuben Glass)

217

And the stories of Crete had hovered too. Had Zeus been born there? Was Minos ruler of the sea? Had Athenian girls truly been sacrificed to the Minotaur? Schliemann set about to find out, but unsuccessfully. Political difficulties constantly stood in his way in the excavation of Crete. No, it would not be Schliemann the dreamer, but another dreamer, Arthur Evans, trained in antiquarian scholarship and, unlike Schliemann, born wealthy, who would succeed. He, too, was driven, particularly by the seals and stones that bore a strange, obscure script. This Minoan language, was it a language far different from Greek? One could only dig to find out. He simply purchased the site that he wanted to excavate—the site of Knossos, the great palace, the very area which Schliemann himself felt should be excavated. In 1900 Evans began his excavation.

The first season was remarkable, indeed it was layer upon layer of culture, civilization after civilization encased in that one spot. Immediately below the surface there was a great palace and within two weeks there were quantities of treasures to lay before Evans. There were, of course, those clay tablets, hardened by fire, but now almost destroyed, containing the script that so fascinated Evans. Eventually some 1600 of these tablets would be turned up. Evans decided that there were three

forms of script: a very early one of simple picture writing, indeed hieroglyphics, and two cursive scripts—one later to be called Linear A, which
was used in the great days of Knossos probably down to about 1400 B.C.,
when Crete was to be invaded by the Mycenaeans; the other Linear B,
probably used by the Mycenaeans both on Crete and in Greece up to the
time of their own collapse.

As the excavations continued, the archaeological world, indeed the
whole world, was startled by what turned up. In the superb palaces, for
example, there were plumbing and water supplies that were superior to
those of nineteenth-century England. The decorations were remarkable,
indeed worthy of palaces, worthy of a people that loved to dance and
sing, to carve, to paint, to create. Who were these strange persons who
loved beauty so much? Their very insularity allowed them what one
professor described he had found in Crete, "the enchantment of a fairy
world."

As the discovery of Minoan art became better known, sensitive
persons responded to it. Scholars were fluent in what they had so say. Sir
Leonard Wooley called it "the most inspired of the ancient art." A dry
analytical historian simply said, "Here and here alone the human bid for
timelessness was disregarded. It was the most complete acceptance of
the grace of life the world has ever known." And that is what the artwork shows us, an extraordinary grace of life.

218

What was this new way of looking at life that the Cretans so developed? Gone was the intense stylization of the Asiatics and in its place
was not a stereotype but fresh new beauty, a tremendous feeling of movement. Deer would spring along the frieze of vases, cats stalked, people
danced or boxed or tumbled, the women were enchantingly beautiful;
great emphasis was put on their eyes, as large as hens' eggs.

Did the Cretans live there on that island home completely isolated?
Why did we not know more about them before? How did they relate to
the Greek mainland? There were in England in the early 1900s arch-
aeologists who felt that the superb sixth-century Greek artwork had
sprung almost ready-made from some curious creative emergence. But not
Evans. He felt more and more that it could be traced distantly to the
past, to these great Cretans with their creative love of life and art.

Cretan art showed early influences from Egypt and the Orient. The
Cretans loved all forms of art, from the most abstract to the ritualistic.
They were traders, trading extensively as merchant adventurers in the
Levant and yet insular, wonderfully protected by their rocky island. Their
sculpture was extraordinarily naturalistic. The Cretans rejoiced in life,
human life, the depiction of everyday men and women. They rejoiced in

the sensuality of life, in clothes, in ritual, in the movements of bodies. And how they loved to dance, with flounces and gestures and raised arms.

They were primitive, of course. In many ways, the games with the bulls, vaguely similar to contemporary bullfights, showed how closely they were tied to the primitive agricultural world from which they sprang. The myth of the bull has always been strong in Crete. It goes back to the very beginnings of their mythology, to the ancient King Minos, who was supposedly a son of Zeus and Europa. Once Poseidon, made angry by Minos, chose only what a god can do and gave Minos's wife, Pasiphae, a passion for a bull. So was born the Minotaur of mythology, half human, half bull, and a monster all the way. One of the sons of Minos was killed by the Athenians, and Minos laid siege to their city. It was a long siege and Minos called upon Zeus to send a plague to the city. To rid themselves of the plague the Athenians consented to an annual tribute of seven youths and seven maidens to be sent to the Island of Crete, where they would be fed to the Minotaur. Eventually the monster was slain by another legendary character of Crete—Theseus. The Minotaur was kept in an amazing labyrinth made by Daedalus, the most cunning of all the Athenians who had invented the ax and the saw. When Theseus became lost on the labyrinth it was Daedalus who gave Ariadne, the daughter of Minos and the beloved of Theseus, the thread which would lead him out of the labyrinth.

In the childhood mythology of islands there was one greater, more beautiful than any other. It was, of course, the island of Atlantis. Ever since the time of Plato, he who first related the story of Atlantis, the lore and legend of this strange sunken island have attracted both faddists and scientists. Only in our own time, however, has science begun to consider the possibility of a true Atlantis. The fantasies that have sprung up about this strange island are worldwide.

In the very beginning, at the dawn of the gods, the earth was divided into sections and to the great Poseidon was awarded the island of Atlantis. On this beautiful island the tempestuous Poseidon walked often upon a plain, a great green meadow, more beautiful and more fertile than any other in the world. Here Poseidon took a mortal woman to wife, and children were born. By another woman, this time a goddess, Poseidon fathered other children. To these children he bequeathed his enchanted island, dividing it into proper domains. His eldest son was, according to the ancient stories, Atlas, and it is from that god that we get our name for the Atlantic Ocean.

A great city was built on the island, a center of power, industry, and trade. The island had everything, the finest of stones, the rarest of min-

Overhanging crystal wave engraved with a map of the ancient legendary island of Atlantis. The map is flanked by figures of the two races believed to have flourished on Atlantis until a cataclysm engulfed the island and its inhabitants. (Courtesy Steuben Glass)

220 erals, even that mysterious metal called gold. It had great trees, innumerable herds of elephants and some animals which have long since disappeared.

Food was plentiful: grapes and corn, vegetables and meat and strange fruits which gave both drink and anointment to the gods. There was no unhappiness on the island; the inhabitants enjoyed complete satisfaction and at the same time enormous technical skill. The ancient Atlanteans knew how to build not only their famous temples, but great bridges, causeways, canals, gymnasiums, and shrines.

Plato tells us:

They arranged the whole country in the following manner: First of all they bridged over the zones of sea which surrounded the ancient metropolis, and made a passage into and out of the royal palace; and then they began to build the palace in the habitation of the gods and of their ancestors. This they continued to ornament in successive generations, every king surpassing the one who came before him to the utmost of his power, until they made the building a marvel to behold for size and for beauty. And, beginning from the sea, they dug a canal three hundred feet in width and one hundred feet in depth, and fifty stadia in length, which they carried through to the outermost zone, making a passage from the sea up to this, which became a harbor, and leaving an opening sufficient to enable the largest vessels to find ingress. Moreover, they divided the zones of land which parted the zones of sea, constructing bridges of such a width as would leave a passage for a single trireme to pass out of one into another, and roofed them over; and there was a way underneath for the ships, for the banks of the zones were raised considerably above the water. . . . The stone which was used in the work they quarried from underneath the centre island and from underneath the zones, on the outer as well as the inner side. One kind of stone was white, another black, and a third red; and, as they quarried, they at the same time hollowed out docks double within, having roofs formed out of the native rock. Some of their buildings were simple, but in others they put together different stones, which they intermingled for the sake of ornament, to be a natural source of delight. The entire circuit of the wall which went round the outermost one they covered with a coating of brass, and the circuit of the next wall they coated with tin, and the third, which encompassed the citadel, flashed with the red light of orichalcum. The palaces in the interior of the citadel were constructed in this wise: In the centre was a holy temple dedicated to Cleito and Poseidon, which remained inaccessible, and was surrounded by an enclosure of gold; this was the spot in which they originally begat the race of the ten princes, and thither they annually brought the fruits of the earth in their season from all the ten portions, and performed sacrifices to each of them. Here, too, was Poseidon's own temple, of a stadium in length and half a stadium in width, and of a proportionate height, having a sort of barbaric splendor. All the outside of the temple, with the exception of the pinnacles, they covered with silver, and the pinnacles with gold. In the interior of the temple the roof was of ivory, adorned everywhere with gold and silver and orichalcum; all the other parts of the walls and pillars and floor they lined with orichalcum. In the temple they placed statues of gold: there was the god himself standing in a chariot—the charioteer of six winged horses—and of such a size that he touched the roof of the building with his head; around him there were a hundred Nereids riding on dolphins, for such was thought to be the number of them in that day. . . .

221

So here on this golden island lived golden people. But time corrupted the people and perhaps eroded the land. From afar Zeus looked on at the depravement of the people, at the denuding of the land, and he ordered a council of the gods. We do not know what the gods said, but according to the ancients the island disappeared into the sea.

We know, too, that there was an abrupt end to the mighty Minoan empire. Were the legend of Atlantis and the death of the Minoan civilization in some way connected?

There is now real evidence that Crete was hit by a gigantic sea wave in 1470 B.C. following the eruption of the volcano Thera in the Cyclades, which the Minoans controlled. Thera is the ancient name for the volcanic island Santorin that is linked today with the destruction of Atlantis.

There are over fifty of the Cyclades, and they contain more than 1000 square miles. Behind their rugged barriers at the sea are green val-

The Minoan palace of Phaestos on the island of Crete. The large vessels were used to store olive oil. (Courtesy Greek National Tourist Office)

leys and fertile lemon groves. One of them, Naxos, was the island where Byron wanted to spend the remaining years of his life if he was granted them.

After Plato the story of Atlantis then went into popular history, touched once again in the fourth century B.C. by the historian Diodorus Siculus of Sicily, who was a traveler and a compiler of ancient history and geographies. He, too, was interested in islands, particularly that inhabited by the Amazons, the garden of the Hesperides. It was on that island that golden apples grew, guarded by a dragon. It was an island of fruit trees, great flocks of sheep, and a race of warrior women who had made themselves into an army of 30,000 on foot and 2000 on horse, who subjugated villages and towns. Some thought that the Hesperides was the ancient island of Atlantis.

And so they went on, the ancients, some saying that the stories were literary ornaments, others that there was indeed a great island that had disappeared. In the seventeenth century a Swede, Olaus Rudbeck, said that Sweden itself was Atlantis while Jean Bailly, living at the time of Voltaire, also thought it was in some area of the north when that land was as warm as the tropics. Scholars and pseudoscholars went on to identify Atlantis everywhere. Was it not in Asia? Was it not indeed in

America? When Francis Bacon wrote his *Nova Atlantis* he placed it in the Pacific, but almost every island from Ireland to New Zealand was at one time or another identified as the mythical island of Atlantis.

Poets took up the quest. In ancient Irish literature, in the Celtic literature of Britain and particularly in Welsh literature the isle of Atlantis took strange forms. There was, for example, the Isle of Avalon, on which lived a legendary character, Morgan le Fay, the woman of the sea. The foster mother of Lancelot, she was the lady of the lake who healed Arthur.

The Irish in the *Voyage of Maildun* enumerated islands, all of which the old Atlantean aficionados considered "memories" of the ancient tradition of Atlantis. The legends and sometimes half-baked scientific theories continued. There were some who said that Atlantis had been the center, indeed the cradle, of animal life. Even the extraordinary lemming was a seeker, some semiscientists said, for the old island of Atlantis. How else to explain those weird migrations which led them to drown off the

223

Remains of the Greek theater at Syracuse, Sicily. Most of these outdoor theaters were built right into the sides of the surrounding hills. (Courtesy Italian State Tourist Office)

coasts of continents? Certainly, some said, it was because they were looking for the lost Atlantis.

But then entered science, and Atlantis is once again in the news. Excavations are being conducted on the island of Thera, which erupted and all but disappeared about 1500 B.C. Carbon dating rather than guessing, archaeology rather than fantasy, began to take over. But just as with the other island mystery—the deciphering of the Cretan language—there are still controversy and bewilderment about the entire story of Atlantis. Just as Schliemann was haunted by the stories of Troy, contemporary scientists and oceanographers are haunted by the old Atlantean legends. There were sunken cities all over the world, the sunken city of Ys in Brittany, for example. There is a "lost island" tradition among the Delaware Indians; there are, as well, deluge tales of sunken lands in the Caribbean and the islands of Mexico.

Myths of the great Utopian islands have always had a strange appeal to man. It is a natural enough appeal, the desire to seek the perfect land, to evolve an ideal society for our incredibly chaotic world. Sir Thomas More, Lord Chancellor of England, living under the reign of Henry VIII, naturally sought a Utopia, and he called it that—the Island of Utopia. It was a land where the common good was considered the only natural evolvement of human life. Sensuality and intellectuality went together. The people lived there on his island according to their nature. More's island was two hundred miles broad, narrow on both ends, a crescent island with the sea between its horns. It was an island of fifty-four cities, but each man of each city was supplied with what he needed. There was no want; More reminds us that it is the fear of want that makes all of us greedy or ravenous. More's island was not far different from Shakespeare's island of *The Tempest*.

Utopias are indeed fantasies, civilizations that were "lost" because they never existed. But what of some of the "lost" island civilizations that did exist?

In the harshness of volcanic Easter Island questions arise, standing up as bluntly as the famous and mysterious long eared sculptured men of the island. Of all the strange islands of the world, Easter Island is perhaps the strangest. Its ancient legends are those of the great migration, similar to all the Polynesian migration legends throughout the Pacific. In 1888 a paymaster in the U. S. Navy, William Naster, made an effort to study the island; he explains that "the tradition goes back before the advent of the people on the island and states that Hotu Matua and his followers came from the group of islands lying toward the rising sun, and that the name of the island was Marae-Toe-Hau, the literal meaning of which is The Burial Place.

224

Végétaux	Végétaux	Ethnographie	Ethnographie
101 — Uhi tapamea. Igname rouge.	110 — Rau hei. Branche de mimosa. (signe de mort).	121 — Toga e te hukiga. Faîtage et charpente.	132 — Vero. Trait, javelot.
102 — Hipu. Citrouilles.	111 — Ira. Liane.	122 — Hahe. Échafaudage.	133 — Mata. Pumex attachés à un bois.
103 — Kava. Gingembre.	**Ethnographie**	123 — Toga. Colonnes de case.	134 — Hanu. Chapeaux en corde d'hibiscus.
104 — Kua hua te kava. Gingembre en fleur.	112 — Nohoga. Case.	124 — Tino. Quille de canot.	135 — Hoga i te pare. Métier pour chapeaux.
105 — Hua. Fleurs.	113 — Hare pure. Lieu de prières.	125 — Vae. Autres bois.	136 — Hau tea. Filasse d'hibiscus.
106 — Huaga. Fruits.	114 — Tagata i te hare pure. Homme en la maison de prière.	126 — Taheta vai. Bassin.	137 — Pepe. Siège.
107 — Rau. Feuilles.	115 — Rakau tau hia no te hare pure. Bois pour maison de prières.	127 — Hakarau. Gaffe.	138 — Vaivai. Bordage.
108 — Kihikihi. Mousses.	116 — Vaka. Pirogue.	128 — Kupega. Filet.	139 — Kona. Perchoir.
109 — Iua tupu te rakau. Plantes en pousse.	117 — Toki. Haches en pierre.	129 — Hura. Petits filets.	140 — Hipu gutu hua. Gourde goulot en bas.
	118 — Huki hoko, oka. Bêche ou lance.	130 — Izizière. Instrument.	141 — Maro. Plumes.
	119 — Tarahoi. Instrument pour graver, pointe, épine.	131 — Vero. Lances à pumex.	142 — Vaero. Queue d'oiseaux.
	120 — Hoea. Instrument pour tatouer.		143 — Humu i te vae. Tatouage aux jambes.

Easter Island script from a French rendering.

The great spirit Make-Make is supposed to have appeared to him and made it known that a large uninhabited island could be found by steering toward the setting sun. Hotu Matua landed and named the island Te-Pito-te-Henua, or the "Navel of the Deep."

That indefatigable traveler, Jean de La Pérouse, described the mysterious island in one of those superb island-hopping trips that he made in the mideighteenth century: "The Island from its situation outside of all the frequented tracks of navigators, from its being totally destitute of wood and water, and from the manner and mode of life of its inhabitants, who are extremely desirous to receive, but have nothing to offer in return, may prove an extensive field for the speculations of the naturalist and philosopher, but can in no respect engage the attention of the different maritime powers of Europe."

It has been the naturalist, the philosopher and the art historian and archaeologist whose interests have turned to Easter Island. Annexed in 1888 by Chile and lying 2000 miles from the coast of that country and another 1100 miles from Pitcairn Island, the nearest inhabited island to the west, Easter Island is the last outpost of the Polynesian world.

It is an island that seems to have been left out of history and when Jakob Roggeveen landed there on Easter Day, 1722, giving it its name, it went into the legendary traveler's tale type of history. Roggeveen spoke of the great statues, thirty feet in height, some with gigantic red hats, and ears that hung down to their shoulders. Gonzales, Cook, and La Pérouse all visited the island in later years, each fascinated by the enormous sculptures.

It is the bird, together with the long-eared men, that is one of the major art concepts on the island. There was once an old ceremony on the island, a ceremonial dance of birdmen holding eggs in a ritual dance around the slopes of the volcano. The bird was a symbol of rebirth, but the island often seemed to be one of death. Its lost civilization might have been an elaborate system of ritual murder, and as early as the time of La Pérouse, it was thought that the statues were perhaps sacrificial statues.

By the midnineteenth century visitors were more common to Easter Island and in 1864 tablets were discovered called by the Easter Islanders *kohau rongo-rongo* or the "singing wood." On these tablets—some as long as six feet, some made from laurel and myrtle, but all of driftwood— were strange hieroglyphics, probably carved with catfish teeth. The rongo-rongos were sacred tablets composed by "message" men, perhaps several hundred of them, and they were chanted at regular ceremonies, the art of rongo-rongo being handed down from father to son. These tablets, too, had stories relating always to death and then to rebirth. The secret of the tablets was jealously guarded from any visiting anthropologist, and expeditions to the island were often made in the efforts to decipher the glyphs. Elaborate theories have been worked out as to what these glyphs most nearly represent, relating them to the hieroglyphics of Egypt perhaps, or to the notations of the Mayans, but to this day no one really knows.

Nor do we know the complete answers to the mystery of the statues. Probably made in the cradle of the volcano, many of them weighing thousands of pounds, how where they placed along the shores? Even Captain Cook queried their transport:

> We could hardly conceive how these islanders, wholly unacquainted with any mechanical power, could raise such stupendous figures . . . they must have been a work of immense time, and sufficiently show the ingenuity and perseverance of the islanders in the age in which they were built; for the present inhabitants have most certainly had no hand in them as they do not even repair the foundations of those which are going to decay.

The natives had their own stories as to how these statues were transported. Until recently they would say that "the great magician used to move them with the words of his mouth."

Says tradition, one such image, four tons of it, in the British Museum required 300 sailors and 200 natives to move it a distance of two or three miles. Recently in front of the Seagram Building in New York far from the volcanic island that created it, workmen labored assiduously to raise an Easter Island statue for the other islanders of Manhattan to see. It was an eerie sight—contemplating a lost civilization within the crowds of another island civilization that could well disappear.

13

To Be Cast Upon a Desert Island

PERHAPS best of all are the islands on paper. Those are the ones we see most clearly, have visited most frequently, have walked and mapped—inhabited islands where there are friends to greet us over and over again, treasure to await us, old pirates to terrorize us. And then there is perhaps the most glorious map of the imagination, the desert island, the island of solitude, the golden Eden where we are self-contained, self-important and self-satisfied.

The deep appeal of the island in literature—and particularly the desert island—has never been surpassed since the discovery of Crusoe's island. On April 25, 1719, Daniel Defoe, sixty years old and in desperate financial straits, published the *Life and Strange Surprising Adventures of Robinson Crusoe, of York, Mariner . . .* written by himself.

Poor Defoe was a hard worker. He had to be; his debts were too heavy. They called him then "an animated machine living down a scandalous past." He was soon hounded not only by his debts and his past, but even by his family, being in that instance rather sweetly hounded by his daughter Hannah, who in the year of Crusoe's writing needed a dowry. She got the dowry, but she never did get a husband. Still, in Defoe's effort to give money that would marry his daughter off, she was, as Walter de la Mare said, "the foster mother of the naturalistic novel."

Hounded at home, unappreciated by his audience, he was to publish seven other books in the same year he published *Robinson Crusoe*. Poor Defoe worked away and it certainly is not surprising that one of his comforts would be a desert island. Even for him that must have been glorious: to be cast up; to carve his own livelihood from the natural world the way he had carved it from words; to find an old retainer, a Friday, who would give him all the companionship he needed.

Yet Defoe has Crusoe say when the latter was putting his story on paper, "I enjoy much more solitude in the middle of the greatest collection of mankind in the world, I mean, at London while I am writing this, than ever I could say I delighted in eight and twenty years' confinement to a desolate island." Crusoe's is the story of a glorious solitude, imaginary, successful, the world cut off in the way you want it to be, the needs supplied, the land rarely hostile, but giving and, with hard work, fruitful. If Defoe's Robinson Crusoe had an almost ideal island, his story was modeled on Alexander Selkirk's being cast away on Mas a Tierra, a not so ideal islet in the Juan Fernandez Islands. Occasionally island lovers prefer to locate this island elsewhere; some, for example, say it is Tobago, but need it matter? Crusoe's island is one of the imagination.

It was a time of islands; they were being discovered all over the world and were exciting the readers of diaries, letters, and reports from early mariners. Crusoe's island had been twice landed on by Capt. William Dampier. On the first trip he discovered a marooned Indian on the island who lived not unlike Crusoe himself. He had a gun, a knife, and a little powder, but he made his own harpoons and lances, built himself a wood hut lined with goatskin, hunted the goats and the seals of the territory. Originally he had worn English clothes, but eventually a hairy goatskin became his only covering. He was the first perhaps to fire Defoe's imagination as he did other readers of Dampier, but it was not until January of 1709 that the true hermit, the character Alexander Selcraig, otherwise known as Selkirk, was discovered wilder than a goat on the same island.

A Scotsman, he was born at a seaside village in Fife. He stands there now in stone gazing into some mysterious sky, into solitude of his own. Like many lads of the community, he had difficulty adjusting himself to the rigid demands of the kirk and turned his thoughts to the sea where he would be safe from the strong conformity demanded of him. At nineteen he shipped to sea. He returned once after six years, but still could not adjust, quarreled with his family, and shipped out again in 1703 under the same Captain Dampier, master of the *St. George*, that accompanied Selkirk's ship. They arrived at Juan Fernandez, the ship a miserable hulk, leaking, reeling, and food atrocious, the discipline unbearable.

For Selkirk this was worse than living in Fife. He demanded that he be put ashore and he was marooned on his island in September, 1704. He left us a catalog of his belongings: cats, some clothes, a sea chest, bedding, one pound of gunpowder, a firelock, a bag of bullets, a flint, steel, a bit of tobacco, a hatchet, a knife, mathematical instruments, a kettle, that book of books—the Bible—and a few other books of simple devotions. The first months were horrible; he wrote he was "scarce able to refrain from doing himself violence."

He sat there by the sea, unable to experience anything but a terrible apathy. After a while this terrible depression lifted. He tried to come to terms with his island. He began to reseek his identity. First he must cut his name in a tree to see who he was, then he must learn to feed himself. There were turtles to be had and he ate them. There was wood with which he could make a hut. He found allspice. He made a decent fire. Dampier in his first voyage had planted turnips, which Selkirk was able to locate and eat. Always there were the goats. He began to move so rapidly, to run so fast that he was easily able to capture the goats, and he earmarked many of them which thirty years afterward were identified by Lord Anson.

He made himself clothing, patiently unraveling worn-out garments to get thread; the cats played with the thread. He had bred the cats carefully because the rats were omnipresent on the island.

These, then, were the facts behind Defoe's book. Several years separated Selkirk's experience and Defoe's publication. But between the bare facts, for example, of Selkirk's clothing and this glorious descriptive paragraph of Defoe's there is all the difference in the world:

> I had a great high shapeless cap, made of a goat's skin with a flap hanging down behind. . . . I had a short jacket of goatskin, the skirts coming down to about the middle of my thighs; and a pair of open-kneed breeches of the same. Stockings and shoes I had none, but had made me a pair of . . . buskins, to flap over my legs, and lace on either side like spatterdashes. . . . Under my left arm hung two pouches, both made of goat's skin too; in one of which hung my powder, in the other my shot. At my back I carried my basket, on my shoulder my gun, and over my head, a great clumsy ugly goat-skin umbrella, but which, after all, was the most necessary thing I had about me, next to my gun.

The book, of course, was an immediate success and has remained so till this day, engendering other books of a like nature. There were of course *The Swiss Family Robinson* and later *Masterman Ready* by Capt. Frederick Marryat, and then Ballantyne's *The Coral Island*, then that

230

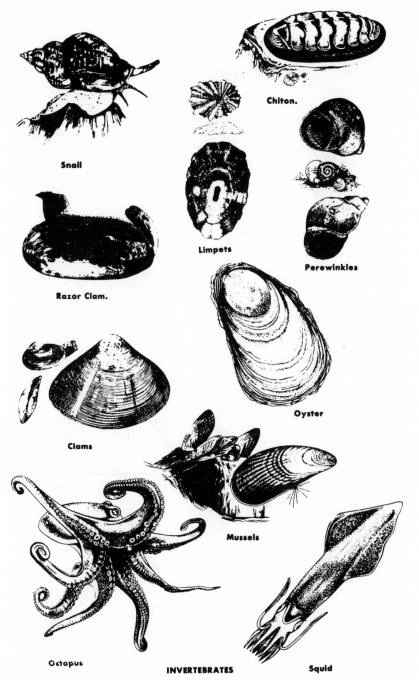

Snail

Chiton.

Limpets

Perewinkles

Razor Clam.

Clams

Oyster

Mussels

Octopus

INVERTEBRATES

Squid

231

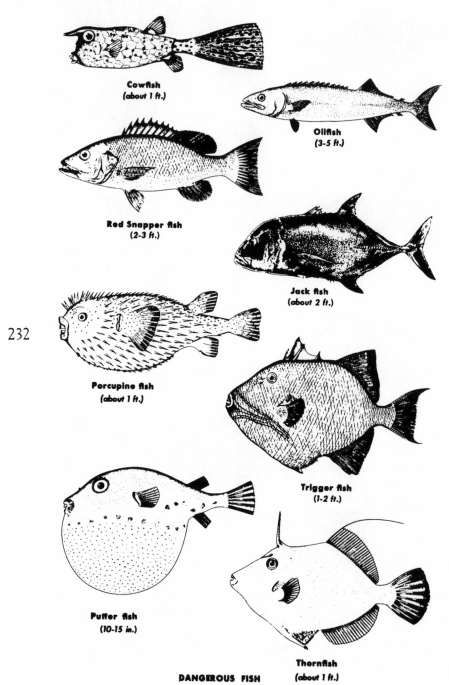

Cowfish
(about 1 ft.)

Oilfish
(3-5 ft.)

Red Snapper fish
(2-3 ft.)

Jack fish
(about 2 ft.)

232

Porcupine fish
(about 1 ft.)

Trigger fish
(1-2 ft.)

Puffer fish
(10-15 in.)

DANGEROUS FISH

Thornfish
(about 1 ft.)

treasure of all, *Treasure Island*, as well as Jules Verne's *Mysterious Island*.

The Swiss Family Robinson is a gentler *Robinson Crusoe*. Who can be lonely with that Robinson family all inhabiting a golden Eden? The history of the book is interesting, as Johann David Wyss, a Swiss clergyman, wrote it—as indeed so many things have been written—for the pleasure of his children. The manuscript was found later by his son Johann Rudolf, a philosophy professor, who edited it for publication. It was a noisy party that landed on that family island, rather different from the ghastly solitude of Crusoe's, and it was an island of conversation:

. . . We next entered a forest to the right, and soon observed that some of the trees were of a singular kind. Fritz, whose sharp eye was continually on a journey of discovery, went up to examine them closely.

"O heavens! father, what odd trees, with wens growing all about their trunks!"

I had soon the surprise and satisfaction of assuring him that they were bottle gourds, the trunks of which bear fruit. "Try to get down one of them, and we will examine it minutely."

"I have got one," cried Fritz, "and it is exactly like a gourd, only the rind is thicker and harder."

"It then, like the rind of that fruit, can be used for making various utensils," observed I; "plates, dishes, basins, flasks. We will give it the name of the gourd-tree."

Then, of course, there was *Treasure Island*, and the story of its writing is nearly as attractive as the story itself:

There was a schoolboy* in the Late Miss McGregor's Cottage, home from the holidays, and much in want of "something craggy to break his mind upon." He had no thought of literature; it was the art of Raphael that received his fleeting suffrages; and with the aid of pen and ink and a shilling box of water colors, he had soon turned one of the rooms into a picture gallery. My more immediate duty toward the gallery was to be showman; but I would sometimes unbend a little, join the artist (so to speak) at the easel, and pass the afternoon with him in a generous emulation, making colored drawings. On one of these occasions, I made the map of an island; it was elaborately and (I thought) beautifully colored; the shape of it took my fancy beyond expression; it contained harbors that pleased me like sonnets; and with the unconsciousness of the predestined, I ticketed my performance "Treasure Island." I am told there are people who do not care for maps, and find it hard to believe. The names, the shapes of the woodlands, the courses of the roads and rivers, the prehistoric footsteps of man still distinctly traceable up hill and down dale, the mills and the ruins, the ponds and the ferries, perhaps the Standing Stone or the Druidic Circle

233

* The stepson of Robert Louis Stevenson.

on the heath; here is an inexhaustible fund of interest for any man with
eyes to see or twopence worth of imagination to understand with. No child
but must remember laying his head in the grass, staring into the infinitesimal
forest and seeing it grow populous with fairy armies. Somewhat in this way,
as I paused upon my map of "Treasure Island," the future character of the
book began to appear there visibly among imaginary woods; and their brown
faces and bright weapons peeped out upon me from unexpected quarters,
as they passed to and fro, fighting and hunting treasure, on these few square
inches of a flat projection. The next thing I knew I had some papers before
me and was writing out a list of chapters. How often have I done so, and
the thing gone no further! But there seemed elements of success about this
enterprise. It was to be a story for boys; no need of psychology or fine writing; and I had a boy at hand to be a touchstone. Women were excluded.
I was unable to handle a brig (which the *Hispaniola* should have been),
but I thought I could make shift to sail her as a schooner without public
shame. And then I had an idea for John Silver from which I promised
myself funds of entertainment; to take an admired friend of mine (whom
the reader very likely knows and admires as much as I do), to deprive him
of all his finer qualities and higher graces of temperament, to leave him with
nothing but his strength, his courage, his quickness, and his magnificent
geniality, and to try to express these in terms of the culture of a raw tar-
paulin. Such psychical surgery is, I think, a common way of "making character;" perhaps it is, indeed, the only way. We can put in the quaint figure
that spoke a hundred words with us yesterday by the wayside; but do we
know him? Our friend with his infinite variety and flexibility, we know—
but can we put him in? Upon the first, we must engraft secondary and
imaginary qualities, possibly all wrong; from the second, knife in hand, we
must cut away and deduct the needless arborescence of his nature, but the
trunk and the few branches that remain we may at least be fairly sure of.

234

On a chill September morning, by the cheek of a brisk fire, and the rain
drumming on the window, I began *The Sea Cook*, for that was the original
title. I have begun (and finished) a number of other books, but I cannot
remember to have sat down to one of them with more complacency. It is
not to be wondered at, for stolen waters are proverbially sweet. I am now
upon a painful chapter. No doubt the parrot once belonged to Robinson
Crusoe. No doubt the skeleton is conveyed from Poe. I think little of these,
they are trifles and details; and no man can hope to have a monopoly of
skeletons or make a corner in talking birds. The stockade, I am told, is from
Masterman Ready. It may be, I care not a jot. These useful writers had
fulfilled the poet's saying: departing, they had left behind them Footprints
on the sands of time, Footprints which perhaps another—and I was the
other! It is my debt to Washington Irving that exercises my conscience,
and justly so, for I believe plagiarism was rarely carried farther. I chanced
to pick up the *Tales of a Traveller* some years ago with a view to an anthol-
ogy of prose narrative, and the book flew up and struck me: Billy Bones, his

chest, and the company in the parlor, the whole inner spirit, and a good deal of the material detail of my first chapters—all were there, all were the property of Washington Irving. But I had no guess of it then as I sat writing by the fireside, in what seemed the spring-tides of a somewhat pedestrian inspiration; nor yet day by day, after lunch, as I read aloud my morning's work to the family. It seemed to me original as sin; it seemed to belong to me like my right eye. I had counted on one boy, I found I had two in my audience. My father caught fire at once with all the romance and childishness of his original nature. . . .

In America, of course, there are islands that inspired literature—Sullivan's Island off the coast of South Carolina, which was the setting for Edgar Allan Poe's *The Gold Bug*, the very name coming from the strange beetle peculiar to the island. But of all the islands only Crusoe's was ever able to inspire this ecstatic description of a true island lover—Dr. L. P. Jack's *Eulogy to an Island*—the most extraordinary piece of island mania on record.

Robinson Crusoe was the first book I read; nor have I ever read another with faith so complete, with imagination so on fire. The sources of thought were tapped; the waters of fancy were unsealed, and the channel cut in which they are doomed to flow until they are lost for ever in the sea. Like a stone dropped into the mouth of a geyser, the reading of that book let loose the floods that boil around the central fires; and a way was made for spirits that haunt the secret springs of life to come and go from that day to this.

235

No philosopher has ever had a clearer conception of the true end of man than I had at the age of twelve. All forms of self-realisation were false save one; and that was, to get oneself cast away, by hook or by crook, upon a Desolate Island. Nothing else would satisfy. Let others go to Heaven if they would; let others be good, or great; but let me be cast on some lonely palm-strewn shore in the uttermost parts of the earth. It was the foolish ship that came to port; it was the wise ship that was wrecked. Not for all the kingdoms of this world would I have exchanged my keg of powder, my cap of goatskin, my fortification, and my raft.

Fundamentally I have never changed that creed. All has been of a piece; there has been no breach in my continuity; the child was father of the man, and my old madness, if madness it was, is with me yet. Before I was twelve years old my fate was sealed; and the gods have kept their bond. It was written that I should explore no Great Continents nor lift up a voice in any City; that the call should come from afar; that I should shun the mainlands where men grow fat and live at ease; that I should stand out into great waters, follow the albatross in her lonely flight, and dwell on the sounding rocks where she makes her nest. O ye Universal Histories and Views of the World, O ye Perfect Satisfactions and Summum

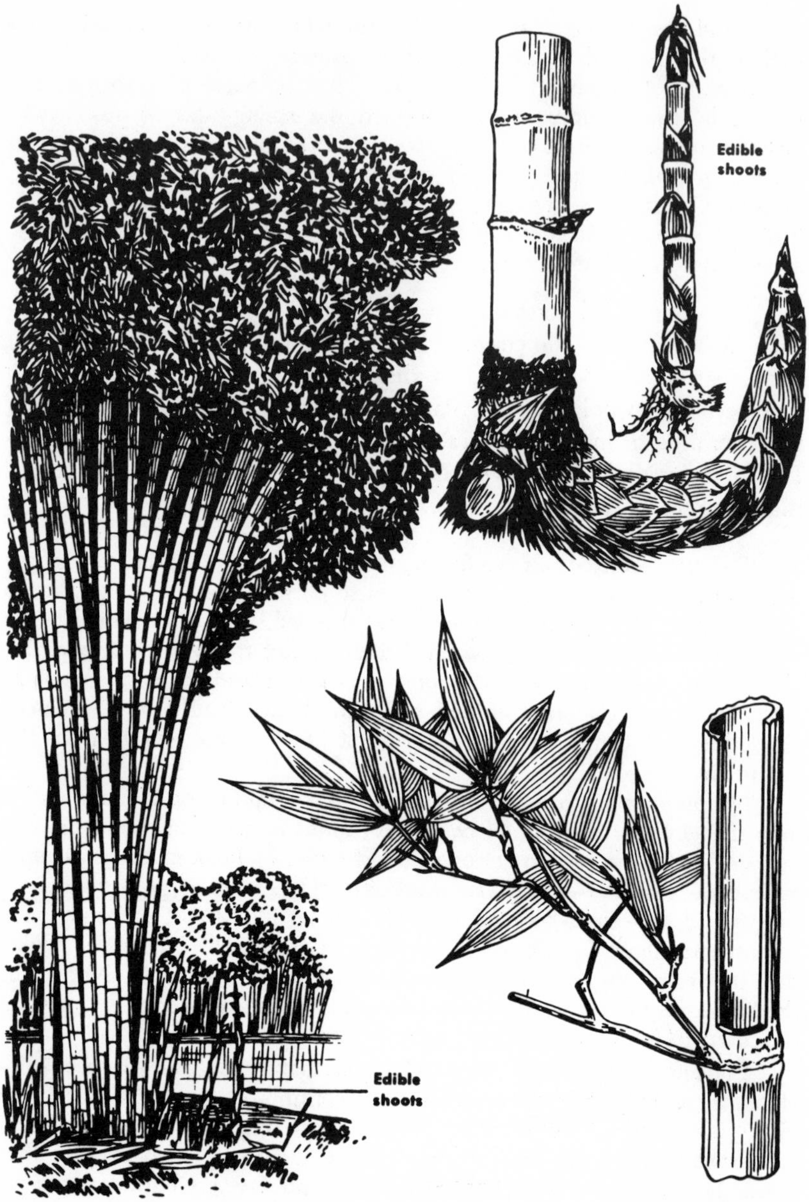

236

Edible
shoots

Edible
shoots

Hollow stem,
use as water vessel

BAMBOO

PALM CABBAGE

Drain sap into
Bamboo joint

GERMINATING NUT

237

Ripe Nut

Husk

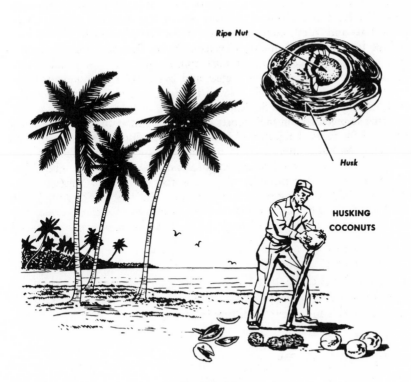

**HUSKING
COCONUTS**

COCONUT PALM

Bonums ready-made, what have I to do with you? A pin-point of time in the wide wastes of eternity, whereon some god had flung me, was all the dwelling-place I ever had. Of broad cities paved with gold I have had no vision; yet have I seen and handled many a shining grain washed down by the River of Life. Good in its Totality I have not known; but hands unseen, held out to save me, have drawn the spent swimmer from the billows of death. Standpoint for viewing, the universe has never been mine; but often, when sinking in deep waters, I have felt a sudden standpoint beneath my feet. With Men in the Mass I have had no traffic. None but lonely souls have I ever met; all were exceptions; Desolate Islands in Time; and with no chart to guide me I have sailed among them, sounding as I went.

Wordsworth speaks of a far-off day when the cataract haunted him like a passion. My haunting passion was the Island. I ransacked libraries for the literature of Islands, and the more desolate they were the better I was pleased. I pored over great maps till Polynesia and Melanesia were more familiar than the geography of the county in which I lived. I found that men who had written of Utopia and other impossible things were as mad as I was about Islands, and I loved them all and read their books over and over again. I knew the Hebrides by heart, I was at home in the archipelagos of the Pacific, I could thread my way among the smallest groups of the Indies, East and West, and a navigator of the Cyclades might almost have used me for a pilot. Columbus, Magellan, Drake, Dampier, Anson, Cook— these were the names of my familiar spirits; and had I not sailed with Odysseus of many devices over leagues and leagues of the unharvested sea? It was always the little islands I loved the best, and if they were not only small but very remote, like St. Kilda, Kerguelen, or Juan Fernandez, so that a mariner shipwrecked on their shores might have reasonable chance of being unrescued for years, I rejoiced like the man who discovered a treasure hid in a field. Australia interested me not the least—it was too big. No castaway of twelve years could be expected to manage such a place. The Channel Islands were contemptible; they were too near. They suggested the odious possibility of being rescued by a steamer! But the Isles of Aru, Tinian and Tidore, the Dampier Group, the Solomons, the Celebes—these were the places where a castaway of merit might make his mark. . . .

Most of us are unlikely to be cast upon such islands, but all of us have had fantasies about them. There is also a stark reality recognized particularly by the U.S. Air Force. Within every Survival Kit there are guides to surviving on tropical islands. We need not chase after Crusoe's goats when we can benefit from the following advice:

Sea Food

Coral rocks, along beaches or extending out into deeper water as reefs, provide the greatest amount of survival food. The more exposed

Daniel Defoe.

239

"I stood like one thunderstruck."
(Robinson Crusoe)

"I descended a little on the side of that delicious valley." (Robinson Crusoe)

surface of the reef bears clinging shellfish. Be sure that all the shellfish you take are healthy. Do not select them from colonies where some are dead or dying.

Fish, crabs, lobsters, crayfish, sea urchins, and small octopi can be poked out of holes, crevices, or rock pools. Be ready to spear them before they move off into deep water. If they are in deeper water, tease them shoreward with a baited hook, piece of string, or stick. You will find flowerlike sea anemones in pools and crevices. They shrink closed when you touch them. Detach them with a knife. Wash well to remove slime and dirt in and outside of animal; boil or simmer.

A small heap of empty oyster shells near a hole may indicate an octopus. Drop a baited hook into the hole and wait until the octopus has entirely surrounded the hook and line; then lift it up quickly. To kill, pierce it with your fish spear. Octopi are not scavengers, like sharks, but hunters, fond of spiny lobster and other crablike fish. At night they come into shallow water and can then be easily seen and speared.

Snails and limpets cling to rocks and seaweed from the low water mark up. Large snails called chitons adhere tightly to rocks just above the surf line. Mussels usually form dense colonies in rock pools, on logs, or at the base of boulders. Black mussels are poisonous in tropical zones during the summer, especially when seas are highly phosphorescent.

240   Sluggish sea cucumbers and conchs (large snails) live in deep water. The sea cucumber can and does shoot out his stomach when excited. Don't eat that; boil the animal and eat the five strips of muscle inside the body and the skin. It will make a gelatinous soup.

TROPICAL FISH

Fish with Poisonous Flesh

There are no simple rules for telling undesirable fish from the desirable ones. Often those considered edible in one locality may be unwholesome elsewhere, depending on the place, their food, or even the season of the year. Cooking does not destroy the poison.

Large barracudas can cause serious digestive illness; yet those less than 3 feet long have been eaten with safety. The oilfish has a white, flaky, rather tasty flesh which is very poisonous. This fish of the Southwest Pacific and all great sea eels should be carefully avoided. Never eat entrails or eggs of any tropical fish.

Undesirable fish have certain characteristics:

Almost all live in shallow waters of lagoons or reefs.

Almost all have round or box-like bodies with hard, shell-like skins covered with bony plates or spines. They have small, parrot-like mouths, small gill openings; and the belly fins are small or absent. Their names suggest their shapes—puller fish, file fish, globe fish, trigger fish, trunk fish.

Fish and Shellfish with Venomous Spines

Reefs are no place for bare feet. Coral, dead or alive, can cut them to ribbons. Seemingly harmless sponges and sea urchins can slip fine needles of lime or silica into your skin, and they will break off and fester. Don't dig them out; use lime juice, if available, to dissolve them. The almost invisible stonefish will not move from your path. It has 13 poisoned spines that will cause you agony and death. Treat as for snakebite.

Don't probe with your hands into dark holes; use a stick. Don't step freely over muddy or sandy bottoms of rivers and seashores; slide your feet along the bottom. In this way you will avoid stepping on sting rays or other sharp-spined animals. If you step on a sting ray, you push its body down, giving it leverage to throw its tail up and stab you with its stinging spine. A sting ray's broken-off spine can be removed only by cutting it out.

Cone shell and long, slender, pointed terebra snails have poison teeth and can bite. Cone snails have smooth, colorful mottled shells with elongate, narrow openings. They live under rocks, in crevices of coral reefs, and along rock shores of protected bays. They are shy and are most active at night. They have a long mouth and a snout or proboscis which is used to jab or inject their teeth. These teeth are actually tiny hypodermic needles, with a tiny poison gland on the back end of each. This action is swift, producing acute pain, swelling, paralysis, blindness, and possible death in four hours. Avoid handling all cone snails.

241

Ferocious Fish

In crossing deeper portions of a reef, check the reef edge shadows for sharks, barracudas, and moray eels. Morays are angry, vicious, and aggressive when disturbed. They hide in dark holes among the reefs.

In salt water estuaries, bays, or lagoons, man-eating sharks may come in very close to shore. Many sharks have attacked in shallow water on bathing beaches in the tropic seas. Barracudas have also made such attacks. Usually sharks four feet long and shorter are timid. Beware, however, of all larger ones, including hammerheads. They are potentially dangerous. not all sharks show fins above the water. . . .

Where to Find Tropical Plant Foods

Tropical food plants will occur most abundantly in open forest clearings, abandoned by natives; along the seashore and margins of streams, and in swamps. The wet, dense jungles are the poorest place to look for survival food.

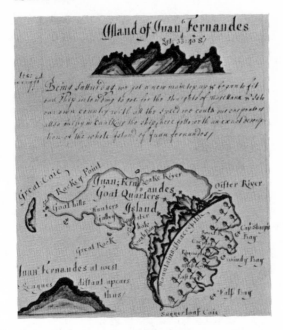

242　The Island of Juan Fernandez. (From *Bartholomew Sharp's Journal*)

The best place to find food plants is an abandoned native garden. In many parts of the tropics, people live in small isolated villages and grow their food in nearby gardens or small clearings. If you find cultivated gardens in the forest, watch out for hostile natives who may be working or guarding the fields; find the trail that leads from the garden to the village. You may find yourself in enemy territory where you may want to take some food but avoid the owners.

Look first for fruits, nuts, or seeds. They can be used immediately for food. The tender end buds or starchy centers of some palm trunks, young bamboo shoots, roots of grasses, and the shoots and flower buds of the wild banana are all good sources of food. Ferns are usually abundant in the moist tropics and make good greens. Even when no food is available, the tender twigs of many plants may be chewed; most kinds have some food value. . . .

Bamboo (Many kinds)

Where found: Bamboos occur most abundantly in the moist temperate and tropical regions. Bamboos are predominantly forest plants.

Habit: The jointed stems (culms) of the bamboo distinguish this plant as a kind of grass. The bamboos are the trees of the grass family. The smallest kinds of bamboo resemble swamp grass, but the largest kinds may develop stems 120 feet high and a foot in diameter.

Alexander Selkirk brought on board.

243

Alexander Selkirk left on the island.

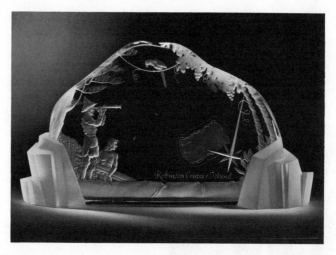

Bower-like crystal form engraved with figures of Daniel Defoe's Robinson Crusoe and his man Friday, and cut at its base to suggest a rocky shore. Crusoe scans the horizon for a passing ship. A map of his island appears in the distance. (Courtesy Steuben Glass)

What to eat: The young shoots of bamboo are edible and appear in quantity during and immediately following rains. They grow very rapidly, some as much as 15 inches a day. But like other wild plants, the edible qualities of bamboo shoots vary. All kinds should be boiled to remove the bitterness; a second boiling in a water change may even be necessary. Some kinds must be buried in mud for 3–4 days to remove the bitterness. Bamboo shoots may be salted, raw or boiled, and eaten as a pickle; they have as much food value as asparagus.

Seeds: The grain of the flowering bamboo may be eaten. Pulverize and with a little water press into cakes or boil as you would rice.

Sugar Cane

Sugar cane occurs widely throughout the tropics. The cane roughly resembles corn with yellow, green, or reddish stems; its leaves are near the top, but "ears" do not develop. The outer layer of the stem may be peeled off and the inside pith chewed for the refreshing and nourishing sweet sap. . . .

Palms

Where found: At least 1,500 different kinds of palms are distributed throughout the tropical world. They grow in almost every conceivable habitat—seashore, swamp, desert, grassland, and jungle. Palms vary in size from a few feet to 100 feet tall. Some are climbers, such as the rattan palms. The palms assume many different forms, but generally they are easy to recognize. The leaves are of two main types: pinnate (like a feather), such as the date palm, or palmate (like a hand with webbed fingers), such as the fan or cabbage palm.

What to eat: Cabbage. The cabbage (terminal bud) or growing point of most palms is edible either cooked or raw. It is located on the tip of the trunk, often rather deeply buried, but enclosed by the crown of leaves of sheathing bases of the leaf stem. Some, but not all, cabbages are bitter.

Sap: The sap of many palms is drinkable and nourishing.

Fruits: The nuts of palms are generally produced in clusters below the leafy crown. Nuts of all New World palms are edible, although many are woody and, therefore, unpalatable. None are poisonous. Nuts of several Old World palms—fishtail and sugar palms—contain microscopic stinging crystals which cause immediate intense pain if eaten. But the fruits of most Old World palms are edible, if not too woody.

Starch: Enormous quantities of edible starch are stored in the trunks of the sago, sugar, fishtail, and giant buri palms. These palms occur principally in southeast Asia and neighboring islands of Indonesia. Another plant, the cycad, found throughout the same area, produces quantities of starch from its thick trunk. The palm-like cycad looks like a cross between a tree fern and a palm.

The pith of the sago palm is used for food in the Southwest Pacific and in Southeast Asia. Cut the tree down before flowers appear; then remove the outer bark, revealing the inner pith. Mash or knead it in a trough made from the base of a sago stem. Let the starch water run into another sago container, where it will precipitate into a fine flour. Pour off excess water. Cook as you would oatmeal, boiling in water until it is thick. Dip out spoonfuls onto leaves and allow them to cool. These gelatinous cakes may be eaten at once or kept for several days. You may also make pancakes of sago, baking them on stone or pottery. Slices of the pith may be baked.

245

Coconut Palm (Cocos nucifera)

Where found: The coconut is widely cultivated and grows wild throughout much of the moist tropics, especially on the east coast of Africa, tropical America, Asia, and the South Pacific islands. It grows mostly near the seashore, but sometimes occurs some distance inland. It does not abound along desert coasts, especially the west coast of continental areas.

What to eat: Cabbage. The cabbage or growing heart is an excellent vegetable cooked or raw. This delicacy has been called "Millionaire's Salad."

Nuts: All or part of the husk of the young nut may be sweet; if so, chew it like cane. Drink the milk from the nut. You may get over two pints of cold fluid from one young nut, especially at the jelly stage, when the flesh is soft. A ripe nut will gurgle when shaken near your ear. But do not drink from very young or old nuts.

Good luck! And take a good book along. It's far better than being stranded on an ice island.

14

Ice Islands

"I HAVE BEEN more than thirty years at sea in almost every kind of situation including torpedo sinkings, fire, hurricane and typhoon, but it is all a new ball game up here." With these words Capt. Robert A. Steward, master of the tanker *Manhattan*, described the epoch-making journey of a vessel becoming the first commercial ship to negotiate the Northwest Passage to Alaska.

The Northwest Passage is a dream more than four hundred years old, a dream that as soon as the New World was discovered made men search for a passage to the fabulous islands of the Indies. It became an obsession that turned man's thoughts to the Arctic and the exploration and development of the different kind of islands—the extraordinary Arctic islands of the north. Their names and their stories are legion: Devon, Cornwallis, Bathurst, Melville, Eglinton, Baffin, Somerset, Prince of Wales, Victoria and Banks; King William Island and Greenland, all of them and more. From the time of the Elizabethans who sailed near their shores, fishermen, whalers, explorers have all been pursued by not only the dream of the Northwest Passage itself, but the dream of opening up and discovering a new world.

The Arctic passage skirted what was called the Unknown Coast. It was approached by small ships; expeditions were decimated by cold, by bears, by disasters of all kinds. From Cabot to Frobisher to Hudson, from Captain James Cook to Parry, the voyagers brought back extraordinary and fabulous stories of almost legendary islands.

Originally it was a race to the other islands. Columbus sailing west from Spain to the Indies; John Cabot, also born in Genoa, refusing to be outdone and deciding that he, too, would search for the islands, the Isles of Spices, by sailing across the Western Ocean—the old name for the Atlantic. In 1496 King Henry VII gave him his formal letters patent:

To our well beloved John Cabot citizen of Venice, to Lewis, Sebastian, and Santius, sonnes of the sayd John, and to the heires of them, and every one of them, and their deputies, full and free authority, leave, and power to saile to all parts, countreys, and seas of the East, of the West and of the North, under our banners and ensignes, with five ships of what burthen or quantity soever they be, and as many mariners or men as they will have with them in the sayd ships, upon their owne proper costs and charges, to seeke out, discover, and finde whatsoever isles, countreys, regions or provinces of the heathen and infidels whatsoever they be, and in what part of the world soever they be, which before this time have bene unknowen to all Christians.

It was a journey that ended up with the discovery of Cape Breton Island, the sighting of Newfoundland, the establishment of a great fishery there, and the reaching of the shores of Greenland in an ensuing voyage.

247

It was, however, a far different thing to moor at Greenland than it was to explore it. It took a couple of hundred years for the necessary skill and dedication to emerge in one man, Fridtjof Nansen, who entered upon one of the greatest island adventures of all, the first crossing of .Greenland.

In the summer of 1882 Nansen was on board the Norwegian sailer the *Viking*, trapped by ice off the east coast of unexplored Greenland. There, for three days, they drifted always closer to the rocky coast where floating ice peaks and glaciers glittering in the daylight stunned their senses. Only the evening heaven with its blazing wild beauty gave them some warmth. Nansen found himself drawn day after day to the deck, and there he turned his glass to the west, "drawn," he said, "irresistibly to the charms and mysteries of this unknown world." For several years the young man brooded and then in January of 1888 he published in Norway his plan to penetrate into the interior of Greenland:

My plan, described briefly, is as follows: With three or four of the best and strongest "skilöbers" I can lay my hands on, I mean to leave Iceland in the beginning of June on board a Norwegian sealer, make for the east coast of Greenland, and try in about lat. 66° N. to get as near to the shore as pos-

248

The crash of the iceberg.

A scene in the Arctic regions.

Greenlanders hunting wild fowl with trident darts. (From Sir Martin Frobisher's *Navigatione in Regiones Occidentis*, printed by Katherina Gerlach, Nuremberg, 1580)

249

sible. I should have liked to land farther north in the unknown regions of Scoresby Fjord, but for this it would be necessary to hire a special vessel, and, as it would probably be difficult to raise funds for this purpose, I have for the present given up this idea. If our vessel is not able to reach the shore, though the sealers, who have often been close in under this unexplored coast, do not consider such a thing improbable, the expedition will leave the ship at the farthest point that can be reached, and will pass over the ice to land. In the summer of 1884, for instance, there was extremely little ice, and the seal were taken almost close under the shore. For the purpose of crossing the open water which will probably be found near the coast, a light boat will be dragged on runners over the ice. That such a crossing of the ice is possible, I feel I can assert with confidence from my previous experience. . . . I should like it to be for preference somewhat north of Cape Dan, where the coast has never yet been explored by Europeans, and offers in itself much of interest to the traveller. To the south the coast is now comparatively well known, as the Danish "Konebaad" expedition,* under Captain Holm, in

* The Danish "Konebaad" expedition to the east coast of Greenland was under the command of Capt. G. Holm and Lt. V. Garde, and was engaged in exploration

1884 reached a point to the north of Cape Dan, and wintered at Angmag-salik, a colony of heathen Eskimo, in the neighborhood of the cape. After having examined the coast as far as the time at our disposal will allow, we shall begin the crossing of the "Inland ice" at the first opportunity. If we reach land to the north of Cape Dan, we shall begin the ascent from the end of one of the fjords close by; if we land farther south, we shall push up to the end of Sermilikfjord before we take to the ice.

We shall try at once to climb as high as possible on the bare rock, even if the gradient be considerably steeper; for, when we are eventually obliged to take to the ice, we shall thus find it flatter and smoother, and shall escape the worst ice-falls of the glaciers, which with their crevasses and general roughness would be likely to prove troublesome and dangerous. Once upon the ice, we shall set our course for Christianshaab, on Disco Bay, and try to reach our destination as soon as possible. . . .

The distance from the point on the east coast where I intend to land to Disco Bay is about 670 kilometres or 420 miles. If we calculate that we shall be able to cover on a daily average from fifteen to twenty miles, which is exceedingly little for a "skilöber," the crossing will not take more than a month, and if we carry with us provisions for double that time there seems to be every probability of our success.

The provisions will have to be hauled on sledges of one kind or another, and besides the "ski" we shall also take "truger," the Norwegian counterpart of the Canadian snowshoe, which may serve our purpose better when the snow is wet and soft. We shall also, of course, take the instruments neces-sary for observations. . . &c. &c.

250

The expedition, which would be successful, was different from all other expeditions, too, because it owed its origin to the Norwegian sport of skiing. Nansen had been skiing since he was four years old and each of the men who accompanied him was also experienced in the sport, about which Nansen wrote poetically:

Of all the sports of Norway, "skilöbning" is the most national and characteristic, and I cannot think that I go too far when I claim for it, as practised in our country, a position in the very first rank of the sports of the world. I know no form of sport which so evenly develops the muscles, which renders the body so strong and elastic, which teaches so well the qualities of

---

from 1883 to 1885. During the summer of 1884 Captain Holm, with the Norwegian geologist Knudsen and a section of his party, penetrated to the hitherto unknown region of Angmagsalik and Cape Dan, where they spent the following winter. "Konebaad," literally "woman-boat," is the Danish equivalent for the Eskimo "umiak," the native skin-boat, which is always manned by women. The expedition here referred to made use of these "umiaks" and Eskimo rowers entirely, as at that time it was considered that the ice-belt of the east coast could only be navigated in this way. (Nansen's footnote.)

Amundsen's ship the *Fram* and Scott's ship the *Terra Nova* at anchor in the Bay of Whales in 1911.

One of Nansen's sail-equipped sledges, south of Cape Richthofen. (1896)

Nansen's first attempts at sailing on snow.

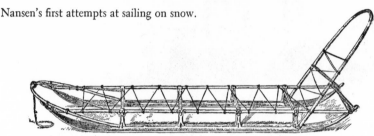

252

Nansen's sledge.

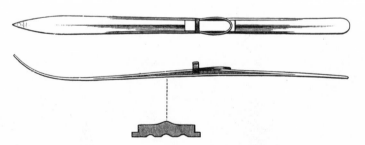

The "ski" of the Nansen expedition in plan, elevation, and section.

dexterity and resource, which in an equal degree calls for decision and resolution, and which gives the same vigour and exhilaration to mind and body alike. . . . There is something in the whole which develops soul and not body alone, and the sport is perhaps of far greater national importance than is generally supposed.

Some skis of course had been used in earlier expeditions, but none with the absolute faith and skill of Nansen.

Even the voyage out was a voyage to islands, first to Iceland, where they would pick up a Norwegian sealer. Bad news reached them in the Faeroe Islands that Iceland itself was packed with ice, but they provisioned there eventually and started out. After suffering in drift ice they finally landed on the east coast of Greenland, met the Eskimos and then began the ascent of the ice.

The going was rough. The cold rasped at their lungs. They moved from one horizon to another without a change of scenery. Everyplace there was whiteness—and flatness. They were alone. Only the sun, the never-ending snow, with no objects to rest the eye upon, no sign by which to direct a course. Although the skiing was difficult, Nansen felt it saved their lives.

I ought perhaps to state once and for all that they were an absolute necessity, that without their help we should have advanced very little way, and even then died miserably or have been compelled to return. I have already said that "ski" are considerably better than Indian snowshoes, even for hauling purposes. They tire one less both because they have not to be lifted, but merely driven forwards, and because the legs are kept no wider apart than in ordinary walking. For nineteen days continuously we used our "ski" from early morning till late in the evening, and the distance we thus covered was not much less than 240 miles.

253

It grew difficult to open the lips to speak—ice formed so heavily on their faces. The biting north wind turned into a whirlwind of snow.

The sky above then cleared completely and it grew colder and colder, the thermometer falling a degree or two below zero. The wind increased in strength; it was bitter work toiling along against it, and we had to be careful not to get badly frozen. First my nose hardened, but I discovered this in time to save it by rubbing it well with snow. I thought myself safe now, but then I felt a queer, chilly feeling under my chin, where I found that my throat was quite numb and stiff. By more rubbing and wrapping some mittens and other things round my neck I put matters straight here. But then came the worst attack of all, as the wind found its way in through my clothes to the region of my stomach and gave rise to horrid pains. . . .

Next morning things were quiet again, but in the afternoon we had another storm of drifting snow from the southwest. This went on all night, the wind working round more and more to the south, and I rejoiced in the hope of a sail, but in the morning again, September 6, it had so far fallen that we did not think it worth while to rig up the sledges. A little later, however, it freshened up and at noon blew due south. I was for sailing, therefore, but the proposal was met with so many objections on the part of the others, who were little inclined for the necessary rigging and lashing in this bitter weather, that I unfortunately gave way. This we all had reason to

254   A Dundee whaler.

regret, for as we went on the wind worked round behind us more and more and at the same time increased in force.

We had soon a full snowstorm blowing from east south-east or east. It was therefore behind us, and carried both the sledges and ourselves on our "ski" along well, and as the ground was also slightly in our favour we made good progress. The driving snow soon grew so dense that Sverdrup and I could not see the others at twenty paces' distance, and we had to wait for them repeatedly in order not to part company. It was no easy matter to get the tent up that evening when we stopped at about eight o'clock, and those unlucky ones among us who had clad themselves insufficiently in the morning and now had to take off their outer clothes to put something extra on beneath had a terrible time. The wind blew in to the very skin; the snow drove through all the pores of shirt and jersey; one felt completely naked in fact; and I myself nearly sacrificed my left hand to the frost in the process, while it was with the greatest difficulty that I could get all buttoned up snug again. The tent we did eventually manage to get up, but we could cook nothing that evening, as the snow drove in much too thick at all crevices and apertures. A few biscuits and some dried meat had to suffice, and we were glad enough to crawl as soon as possible into the sleeping-bags, draw the covers well over our heads, devour our food there, and as we slept leave the storm in undisputed possession outside. We had pushed on a long way that day, not much less than twenty miles, as we supposed. . . .

A stranded Greenland whale.

255

"The ice was here, the ice was there,/The ice was all around;/It cracked and growled, and roared and howled,/Like noises in a swound!"—S. T. Coleridge The *S.S. Manhattan* conquers the Northwest Passage.

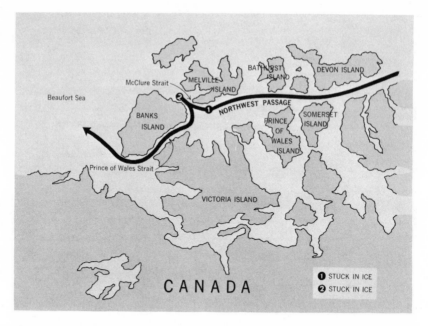

The route of the S.S. *Manhattan*. (Courtesy Humble Oil)

The men began to question,—Ravna's face began to get longer and longer, and one evening about this time he said, "I am an old Lapp, and a silly old fool, too; I don't believe we shall ever get to the coast." I only answered, "That's quite true, Ravna; you are a silly old fool." Whereupon he burst out laughing: "So it's quite true, is it—Ravna is a silly old fool?" and he evidently felt quite consoled by this doubtful compliment. These expressions of anxiety on Ravna's part were very common.

Another day Balto suddenly broke out: "But how on earth can anyone tell how far it is from one side to the other, when no one has been across?"

It was two months since they had beached their boat—but they turned now to the help of an iceboat—a sledge-and-sail arrangement.

Our ship flew over the waves and drifts of snow with a speed that almost took one's breath away. The sledges struggled and groaned, and were strained in every joint as they were whirled over the rough surface, and often indeed they simply jumped from the crest of one wave onto another. I had quite enough to do to hang on behind and keep myself upright on the "ski." . . . Our ship rushed on faster and faster; the snow flew round us and behind us in a cloud, which gradually hid the others from our view.

So with ski and sails the inland ice of Greenland was conquered, but still to be completely understood.

The S.S. *Manhattan*, as of this writing the most powerful tanker under the United States flag, did more than negotiate the ice; the expedition studied it as well. And what was one of their biggest surprises on the voyage among the ice islands of the north?

"The nature of the ice."

The Antarctic and Arctic islands have attracted many adventurers. Their stories make for superb reading. The tourist fortunately is rarely attracted, for as such he, along with the oil seekers, is a threat to the environment. So, alas, is even the scientist as Robert Cushman Murphy has pointed out:

Man has himself become the center of a potential demographic crisis in Antarctica. Proliferation of his occupancy and travel must be accepted. There may be a modest withdrawal during some years, but it appears certain that the remainder of the twentieth century will be mostly years of expansion in his numbers, and that pursuit of scientific information will be the main reason for his presence.

An establishment like McMurdo, the home of a thousand or more human beings in the austral summer and of a couple of hundred in winter, fairly bursts with conservation problems, of which the disposal of sewage, garbage, and other waste is only the first. There is no subsoil seepage, no rapid bacterial decomposition, such as come to our aid in the more familiar world. Each season the biologists find that they must go farther afield for their ecological studies. Even the bottom of Winter Quarters Bay is losing the semblance of a "natural area." . . .

But even such a beautiful "natural area" as the Caribbean is being threatened.

257

15

At Last, the Caribbean

At last we, too, were crossing the Atlantic. At last the dream of forty years, please God, would be fulfilled, and I should see (and happily, not alone) the West Indies and the Spanish Main. From childhood I had studied their Natural History, their charts, their Romances, and alas! their Tragedies; and now, at last, I was about to compare books with facts, and judge for myself of the reported wonders of the Earthly Paradise. We could scarce believe the evidence of our own senses when they told us that we were surely on board a West Indian steamer, and could by no possibility get off it again, save into the ocean, or on the further side of the ocean; and it was not till the morning of the second day, the 3d of December, that we began to be thoroughly aware that we were on the old route of Westward-Ho, and far out in the high seas, while the old world lay behind us like a dream.

So Charles Kingsley began his book (*At Last: A Christmas in the West Indies*) on the West Indies of the nineteenth century. Even today the most world-traveled British who have yet to see the Caribbean can speak of it with the suppressed exuberance of generations of storybook discovery and romance.

The islands are subject now to the last invader: the tourist. But they are used to waves of unknown peoples just as they are used to the waves of the sea. From the mild Arawak to the militant Caribbean (both Indian tribes probably from the great rain forest of South America), the islands have known the gentle wand of cultivation and the hard fist of oppression. The Caribbean is a microcosm of the world; the European power struggle was played out in its blue waters and its sea-grape-girded beaches. They came in caravels with Columbus, in barks, in privateers, in sailing ships from all of Europe—bands of Spanish, French, English, Dutch, Swedes, Danes—they came with dreams and they came with slaves.

The ghosts of other centuries still haunt the Spanish Main: The plantation life of other days haunts the corridors of the Hiltons; the hand of the oppressor is behind every move of the new nationalism of the islands—and the attempt to seek broader economies. The tourist is no more trustworthy than the old buccaneer and sugarcane. It is not that the tourist is bad; he is simply unpredictable . . . and that is partially because of the beauty of the islands. One wants to see them all; in a beautiful family (but a family wracked by dissension) one wants to play no favorites. One year Antigua is "in"; the next year Barbados; the styles change as rapidly as the beach clothes. No, the average tourist is not predictable and he is not particularly observant. He has come for sun, not to look too deeply at the economy (he'll have plenty to do with that word when he gets back Stateside); he has come for the illusion of peace, not to look at the peoples. He wears his ignorance with the same kind of embarrassed pride with which he wears his island straw hat: it's not particularly comfortable, but it protects one from the sharp clarity of the Caribbean sun. He is the new gentle-hearted, affluent conquistador in an Eden just a few hours from home. You can see him a little as Kingsley saw him:

259

Under a cloudless sky, upon a sea, lively yet not unpleasantly rough, we thrashed and leaped along. Ahead of us, one after another, rose high on the southern horizon banks of grey clouds, from under each of which, as we neared it, descended the shoulder of a mighty mountain, dim and grey. Nearer still the grey changed to purple; lowlands rose out of the sea, sloping upwards with those grand and simple concave curves, which betoken, almost always, volcanic land. Nearer still, the purple changed to green. Tall palmtrees and engine-houses stood out against the sky; the surf gleamed white around the base of isolated rocks. A little nearer, and we were under the lee, or western side, of the island. The sea grew smooth as glass; we entered the shade of the island-cloud, and slid along in still unfathomable blue water, close under the shore of what should have been one of the Islands of the Blest.

It was easy, in presence of such scenery, to conceive the exaltation which possessed the souls of the first discoverers of the West Indies. What wonder if they seemed to themselves to have burst into Fairy-Land—to be at the gates of The Earthly Paradise? With such a climate, such a soil, such

Valley in the Blue Mountains, Jamaica. (1887)

vegetation, such fruits, what luxury must not have seemed possible to the dwellers along these shores? What riches too, of gold and jewels, might not be hidden among those forest-shrouded glens and peaks? And beyond, and beyond again, ever new islands, new continents perhaps, an inexhaustible wealth of yet undiscovered worlds.

No wonder that the men rose above themselves, for good and evil; that having, as it seemed to them, found infinitely, they hoped infinitely, and dared infinitely. . . .

No, the tourist does nothing wrong today; it is peace that he is greedy for—and sometimes land, as the land booms throughout the islands reveal. Just as in Florida, where an ignorance of ecological conditions has fostered poor shore projects (to see a thicket of sea grape mowed down for a new condominium because last year's condominium is no longer "in"—"no one buys a secondhand condominium"—is to see the death of a barrier-island coastline) and ill-thought-out public works have caused the most ruinous effect on coastal resources, so the Caribbean islands, too, are threatened.

Blue Basin, Trinidad. (1887)

Will there be any peace for a tourist who cannot, as fortunately we still can—for how long?—leap upon the shore at St. Croix as Kingsley did?:

As we leaped on shore on the white sand, what feelings passed through the heart of at least one of us, who found the dream of forty years translated into facts at last, are best, perhaps, left untold here. But it must be confessed that ere we had stood for two minutes staring at the green wall opposite us, astonishment soon swallowed up, for the time, all other emotions. Astonishment, not at the vast size of anything, for the scrub was not thirty feet high; nor at the gorgeous colours, for very few plants or trees were in flower; but at the wonderful wealth of life. The massiveness, the strangeness, the variety, the very length of the young and still growing shoots was a wonder. We tried, at first in vain, to fix our eyes on some one dominant or typical form, while every form was clamouring, as it were, to be looked at, and a fresh Dryad gazed out of every bush, and with wooing eyes asked to be wooed again. The first two plants, perhaps, we looked steadily at were the Ipomoea pes caprae, lying along the sand in straight shoots thirty feet long, and growing longer, we fancied, while we looked at it, with large bilobed

Port Royal, Jamaica. (1887)

green leaves at every joint, and here and there a great purple convolvulus flower; and next, what we knew at once for the "shore-grape." We had fancied it (and correctly) to be a mere low bushy tree with roundish leaves. But what a bush! with drooping boughs, arched over and through each other, shoots already six feet long, leaves as big as the hand shining like dark velvet, a crimson mid-rib down each, and tiled over each other,—"imbricated," as the botanists would say, in that fashion which gives its peculiar solidity and richness of light and shade to the foliage of an old sycamore; and among these noble shoots and noble leaves, pendent everywhere, long tapering spires of green grapes. This shore-grape, which the West Indians esteem as we might a bramble, we found to be without exception, the most beautiful broad-leaved plant which we had ever seen. Then we admired the Frangipani, a tall and almost leafless shrub with thick fleshy shoots, bearing, in this species, white flowers, which have the fragrance peculiar to certain white blossoms, to the jessamine, the tuberose, the orange, the Gardenia, the night-flowering Cereus; then the Cacti and Aloes; then the first coco-nut with its last year's leaves pale yellow, its new leaves deep green, and its trunk ringing, when struck, like metal; then the sensitive plants; then creeping lianes of a dozen different kinds. Then we shrank back from our first glimpse of a little swamp of foul brown water, backed up by the sand-brush, with trees in every stage of decay, fallen and tangled into a doleful thicket, through which the spider-legged Mangroves rose on stilted roots. . . .

Some islands instantly welcome you. In the old days when the ships pulled up at the wharves, there were of course swimmers or small boats to greet you. There would be a shoreline that you could approach, cautiously moving from the large boat into a small boat, the land growing more welcoming as you stepped on the final dock. There in the blaze of the tropical summer or even in the cool clarity of a Maine morning, you integrated slowly into island life.

You could look around and observe the activity of the wharves. In the north, the fishing boats tied up against the shore, the activity on the dock, quite independent of you. The dock was the spot of livelihood, frequently a panorama of color, always the place of shape and design—horizontals and verticals, different colors in the north, one shading into another with short definition of horizon.

In the tropics it was different. There at first the design confused, with colors blinding the shoreline too sharply and the colors of the houses momentarily too violent to the eye. But now, in our island-hopping world, the plane is our ship. It is bad enough to be interjected into a large country via the airports, but in a small one it is almost a travesty of justice, a shock to the eye, a poor wrench to the heart.

263

A charter sloop sails the trade winds from St. Croix to Buck Island, Christiansted, Virgin Islands. (National Park Service photograph by M. W. Williams)

ISLANDS

All airports, of course, look alike and make all the islands curiously the same. All the colors and shapes are muted, all of the hangars familiar, the planes the same throughout the world, the activity of the airport universal, the same magazines on the racks, the same cigarettes, the same waiting chairs, the same pensive faces, angry hands, strident voices and fatigue—that terrible fatigue of the traveler which puts over his face, for a while, a mask. He is no one going nowhere, knowing no one, his destination—no place; he's caught in an absolute—no time between the place he has left and the place to which he is going.

Sometimes in an airport of such islands, you will see even a more miserable kind of despair, the departing traveler, the other face of the islands—the nontourist face. He is the traveler who is leaving home—perhaps for good, he does not know—leaving the past behind.

That is not true of the Caribbean emigrant. It doesn't matter where he comes from; perhaps he is just island-hopping, too, from one more pathetic island to another, trying to scratch some kind of living, not from the soil but from hands that can wash dishes, or serve the tourist, or open the door, or drive a taxi, or make a drink. The small islands of the Caribbean necklace—they used to call them the "pearls of the Antilles"—are now scarred and scratched through too much traveling. These islands could almost be brothers or sisters, but they are alien to each other, having turned so completely against, or perhaps never having found, the parental or maternal form of government that would support them. They are all siblings, one fighting with the other, each competitive, few with one good word for the next. On the island of which we write now the "aliens" are the islanders from another island just a few miles away. There is no indoctrination for the "alien"—he just comes and then goes away.

We used to be able to see a semblance of indoctrination programs going on in Puerto Rico when, citizen though he was, the Puerto Rican on his way to the States had to be told what to expect. The most important thing, of course, was omitted. The Puerto Rican traveler coming from Coamo or from even farther away in Viecques, a little island off Puerto Rico, would not have the most invaluable of all things, a winter overcoat. In other ports in the islands—in the Virgin Islands for example—there are local jokes as to how long an overcoat can be passed on from one couple to another as they make the pilgrimage back stateside.

This was a humorless situation for the Puerto Rican. His clothes were always too skimpy for the climate to which he was going; the same problem existed for the islanders from the British West Indies who would arrive in London. If anything, their problem was more acute. The Puerto Ricans consider themselves Americans, yes, but Spaniards too. Their culture is of Spain. But in the British islands the natives were more British than the British. Schooled in English schools, starched in English manners, the poor immigrants from the islands would arrive—Indians, blacks, mix-bloods—in the cold fog of an English winter, to find themselves lost in a world they had not made. Some of the Indians found it the most difficult of all; since they came from the island of Trinidad, their culture was Indian, their food was Indian, their spirit was Indian, but they did not belong anywhere, as that wonderful writer V. S. Naipaul pointed out. Prejudiced in two ways, both as black colonials and as Indians, they could express themselves only in their glorious gift for words.

The English-speaking island of Puerto Rico and the French-speaking islands of Martinique and Guadeloupe all have a glorious gift for language. It is the word—not the world—that is the islander's oyster.

So the Caribbean has many faces—and many voices. Many American families *are* tied to the islands with the same intensity of feeling that generations of island lovers have felt. It is still possible to have a very private feeling about your private island—or at least your almost private island—particularly in the small cays of the Bahamas as Carolyn Royal relates (along with some recipes for the glory of the islands—conch):

It's Christmas Day and we are in an airplane over the Bahamas. The 265 children, who never really believed we'd get here ("Something always comes up"), each have a window seat and are rubbernecking and firing questions:

"How deep is the water there, Mum?"

"Deep."

"Why are there all those roads and no houses?" We are flying into Freeport, on Grand Bahamas Island. We make a guess: "Maybe some real estate developer put in the roads and ran out of money before he could build the houses." We learn later they are what New Englanders call "tote roads." Years ago they were built to carry out logs and there the roads were abandoned.

After Kennedy Airport, Freeport Airport looks pretty small. But going through customs and immigration is an adventure, particularly when you're escorted by a uniformed officer of the Queen. Our poodle, Duchess, is watching us from her crate, and barks at us.

Duchess is temporarily out of her crate and on her leash and we visit the International Bazaar. This is pure enchantment! We wander from Spain to China to Mexico on cobblestone walkways, peering in shop windows that display exquisite and tempting goods from many countries. Happily for our wallets most are closed for Christmas.

Conch fritters and conch chowder are our lunch back at the airport. Duchess has returned to her crate and is looking at us with dismay at this further humiliation.

Treasure Cay! The children blink at a one-strip airfield and a minuscule

A limbo dancer demonstrates his skill in Barbados. (Courtesy Air France)

terminal building. But the lure of the Bahamas has already caught them.

A little more flying in our prop-jet and we are in Marsh Harbour, our jumping off place for Man o' War Cay.

Man o' War! That has been a magic name at home in Connecticut since the adult members of the family were there a year before. After twelve months of pleading by the young ones, here we are. But first we must get to our "out island."

Through Air Mail letters sent well in advance, arrangements have been made—but quite unlike the arrangements you would make for a Miami Beach hotel. Victor Russell is there with his taxi, and we drive along the left side of the road, reminding us all again that we are in the United Kingdom. Marcel Albury's "plane boat" is waiting at the dock, and once aboard, we change to sneakers and stow "city shoes" for ten days. Maurice Albury is waiting at the gas dock on Man o' War with the sailing dinghy loaded with enough food to tide us over until the grocery opens tomorrow. There are handshakes and warm greetings each time we meet our friends.

For the first time, the children hear the wonderful way of speaking of the Man o' War Settlement. New Englanders put r's where they don't belong and Southerners drop them. Our Bahamians transpose h's. Thus, "Upper Harbour" becomes "Hupper Arbour." Later, when we're chopping

In the lucent waters of
Nassau in the Bahamas a
native boy dives for conch,
long a staple food here.
(Bahamas News Bureau
photograph by Howard
Glass)

267

open conchs, our son knows what I mean when I ask him to fetch "tuh
haxe."

The forefathers of the folk of the Settlement were Loyalists, or Tories,
from the Carolinas. During the American Revolution their property was
taken away, or taxed so heavily, that many exiled themselves to Canada,
England and the Bahamas, among other places. The Man o' War exiles
settled into what has become a heritage passed down from father to son:
boatbuilding.

Almost everyone here owns or works in a boatyard. Some are in back-
yards, some are more formal. They work from half-models, not blueprints.
The ribs are carved of madeira roots, the planking is Honduran mahogany.
The work is done by hand; you'll seldom see a power tool here. The big boats
are built by "Uncle Will" Albury and Edwin Albury (one of Maurice Albury's sons). They keep the trade going, but many of the youngsters of the
cay are finding living more profitable in Nassau and Florida.

We transfer a huge amount of luggage to the dinghy. No room for us
then, so Mr. Maurice shoves off and we start walking down the Queen's
Highway. We pass Edwin's boatyard and see what's going on since we were
last here. A big one on the ways, another coming along. We're catching up
on what has happened during the last year.

Shipbuilders follow a craft handed down through generations by the seafaring people who live at Man o' War Cay, Abaco. (Courtesy Bahamas News Bureau)

268

La Plata River in the in-
terior of Puerto Rico.
(Courtesy Puerto Rico In-
formation Service)

Haitian voodoo ceremony. The live chicken will be part of the sacrifice to the great god, who is usually represented by a snake. Haitian voodoo is much more structured than the American variety. (Courtesy Air France)

269

We each carry our own flashlight, as it's getting to be the end of Christmas Day. For awhile the Queen's Highway is an asphalt walk (wide enough for two bicycles to pass each other). Shortly it becomes a narrow, sandy path. Later it becomes a beach, and before we get home it's coral rock, then a climb up an obscure path through the underbrush. Along the way we hail friends, then join new friends, the Windsors, who are here for the first time. Lindsay Scott, who owns the home where they are staying, had asked us to introduce them around. No need to: they had an immediate rapport with these friendliest of all friendly people.

We are home. Jim Cole, who owns the house, wonders aloud what kept us. A bit of conversation and making up of beds, and so we end Christmas Day.

A splendid sunrise! I go out while it's still slightly less than night and wait with movie camera and Leica in hand. As the sun comes up (and it seems to come up quicker in the Bahamas than in New England) I grab one camera, then the other. Since the house is on a skinny wing of the cay I can jump from the ocean side to the bay side. Over the ocean the sun rises and the reef becomes orange. Over the bay the low-lying cumulus clouds turn yellow and pink (somewhat reminiscent of a badly tuned color TV set, but we want to forget things like that). The bushes and palm trees are still black, but not for long.

ISLANDS

I'm joined by Duchess, who has staked her claim on certain areas. She romps on the little beach as only a poodle who has springs instead of legs can do. The majesty and silence of the pre-sunrise is suddenly broken. The sun is up, looking as though it had come swimming from the bottom of the ocean. It climbs so rapidly that the movie camera can follow it. The cay is now bright with fresh new sunshine and sixty degree air and a light breeze. The water is again blue and green. A lizard, about as long as my hand and probably the most "dangerous" creature on Man o' War, looks around quizzically and slips into the underbrush to search for breakfast.

Breakfast time for us, as well. No gongs needed. There is a blurred exodus from the cottage of three barefoot, bathing-suited children. The propane stove is lit and coffee water is set to boil. Bacon is on the other burner. Eggs for future scrambling (with onions and the herbs I brought from Connecticut) are in a bowl. Jim meanders out of his room and ponders his corn crop, which has not come up, probably because it's the proverbial watched-pot.

Breakfast is over. Elder son takes the bucket down to the cistern, scrapes the leaves from the rain water there, and we heat a small amount of water for dishwashing. Reminders to all aboard: if you have to conserve fresh water you don't stack dishes. No paper plates allowed; they're horrendously expensive since all provisions are brought in by boat or airplane. Dishwashing over: we discuss which of our two-foot high palm trees is next in line for a dishwater drink.

270

Our children and the Windsor children charge off for a swim. They have snorkels, masks and flippers to use exploring. They come back with a collection of starfish. Despite our effort to dry them on the beach they are pretty stinky in a couple of days.

We work clearing the underbrush to make a bigger garden. The debris is carted down to the beach for a Guy Fawkes bonfire.

The children are tired of the tameness of the bay. They explore the ocean side of the cay and make sand castles, delighting in the fact that they can build one and seconds later a wave removes it.

They play exuberantly, unmindful of the barracudas and undertow. The youngest one, Leif, gets too close to the surf and out he goes. Without knowing what is happening to him he is heading toward the reef, a half mile away. The two older boys are good swimmers. They drag him back to shore. He arrives home, scraped from stem to stern by the sand and coral, and maybe wiser.

All six children are severely lectured and they are reacquainted with the boundaries of their playground.

New Years' Eve is today! Mr. Maurice shows us how to chop open the conchs and I make conch chowder. Anne Windsor bakes chicken and Jim Cole produces a bottle of rum. Anne and I break our week-long tradition of shorts and bathing suits and put on evening slacks. We still wear our sneakers however, and the men and children look the same as they do every day.

Vareck has practiced "Auld Lang Syne" on his harmonica and when we declare it 1969 (at nine p.m.) we listen to a swinging version, followed by "God Save the Queen" and "The Star Spangled Banner."

Home without flashlights. There is a beautiful full moon and the path, which is tricky during the day, is easy to follow tonight. We light one kerosene lantern and talk about old times and old and new friends for a little while. The new year manages to come in without our help.

Every Christmas the Gilbert Bells send a mimeographed list of the names and addresses of the families of the Settlement and the Americans and Canadians who have vacation homes here. As of Christmas, 1968, there were 72 families in the Settlement and 66 from the States and Canada. Of the 72 Bahamian families 33 families are named Albury and 14 are Sands.

Life on Man o' War precludes the possibility of inbreeding that is usual in other insular areas of the world; at any rate, there are no effects of it. There are no nightclubs or bars, not even a restaurant on the cay. Transportation is by foot, bicycle or boat, and a few golf carts in the Settlement. This substantially reduces the tourist invasion.

But there are people who venture here and either turn around dismayed or come back again and again. Those who leave are chagrined at the lack of indoor plumbing and those who come back could not care less. Our friend, Lindsay Scott, and I were sitting on the dock one day in our dirty dungarees

271

In remote parts of Bonaire and Curaçao, drinking water is delivered from house to house by donkey. These islands have very little rainfall, and the well water is not potable. (KLM Royal Dutch Airlines photograph by Fritz Henle)

Antigua's beaches mean mile after mile of sparkling white sand, coconut palms, and turquoise-blue sea. Not far from the water's edge are luxurious hotels which offer fine dining, night life, water sports. (Courtesy Air France)

and were ordered by a "lady" to bring her groceries to her yacht. We responded with overly precise grammar and followed her command, acting like good butlers and snickering all the whole. Good chance she'll never be back to our cay.

Most of the Americans and Canadians order provisions from the two grocery stores by their Citizen Band radios. We prefer to go down in the dinghy, or walk the Queen's Highway. This year we find that Sammy Albury has recently married and now has a good helpmate for running the store. His mother, Miss Mizpah, is on the dock and we sit and chat, learning of deaths and births and marriages.

Another way to get food is to fish for it. Robbie Weatherford picks us up at dawn in his boat. He knows the places to go: the reef, Scotland Cay, the sandbar on the bay side, the shoals. We chum, using the inedible fish you inevitably catch, and we use handlines and rods we brought from Connecticut. The Big One That Got Away is of course the high point of the day. After 45 minutes of battling him our only reward is to see his tail waggling out from under the reef.

The fish we catch have names unfamiliar to us in New England. The most beautiful is the triggerfish and I run out of color film just before he's brought aboard. His color won't last until we get home.

A feast is in store for us. Mr. Robbie supervises the cleaning and we end up with enough food for the ten of us for two suppers and breakfasts. Mr. Robbie says it wasn't a good day.

Last night on Man o' War (for this trip). The moon is waning and the wind is out of the Northeast. We half-hope there will be a hurricane so the plane boat can't take us across to the airport in the morning. We walk down to the water and just look at it. The temptation to have a last wade is overwhelming. Tomorrow's telephones and dictation machines are a far-off probability.

ADDENDA

Conch is pronounced "konk."   Cay is pronounced "key."

Conch Chowder

6 Conchs	salt and pepper
¼ lb. salt pork or bacon	2 qts. water
(finely chopped)	½ c. milk
6 large potatoes (cubed)	(coconut milk if available)
3 large onions (chopped)	1 can tomatoes

Take 6 conchs & either put them through a meat grinder or pound them for a half hour with a hammer; lightly saute ¼ lb. finely chopped salt pork or bacon; cube 6 large potatoes; chop 3 large onions. Put all in large kettle, add salt and pepper, cover with 2 quarts of water. Cook 2½–3 hours, stirring occasionally. Then add ½ c. milk and 1 can of tomatoes. Reheat if necessary.

Added Fillip
(Lindsay's Man O' War Specialty)

Mix (a week or 2 years in advance): 1 bottle inexpensive sherry + 5 hot peppers. Add a dash or two to the chowder.

273

Harbor patrol proceeds at a leisurely pace in Bridgetown, Barbados. (Courtesy Air France)

Conch Fritters

Pound 2 conchs, boil 10 minutes until tender, then put through a meat grinder. Add:

¼ lb. flour	salt
1 tsp. baking powder	a little water or milk
2 eggs	
½ onion, finely chopped	

Mix thoroughly. Fry either in deep fat or like pancakes. Optional: a touch of tomato juice.

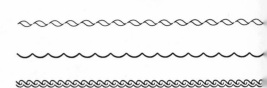

16

Great Islanders

"It was at sunset in Autumn that we were set ashore on that loneliest, lovely rock where the lighthouse looked down on us like some tall black-capped giant and filled me with awe and wonder. At its base a few boats were grouped on the rock standing out dark against the red sky as I looked down at them. The stars were beginning to twinkle; the sound of many waters half bewildered me. Someone began to light the lamps in the tower. Rich red and golden they swung round in mid-air; everything was strange and fascinating and new." So wrote Celia Laighton Thaxter about her arrival on White Island in the Isles of Shoals, whose description she would make famous throughout the world.

She was just four when she arrived on the island, having been unceremoniously uprooted from the quiet little white house in the harbor town of Portsmouth, New Hampshire, to be carried away to an uninhabited island with her parents. There was a cottage there, a little stone cottage with a whitewashed ceiling and deep window seats on which the child could remember sitting and staring out at the fierce and circling sea. She never seemed to feel any isolation; quite the contrary, she felt always at home there, and throughout her life and particularly in later years when she was to become a friend of many great writers who would vacation on the Isles of Shoals, she always returned to find comfort and succor from the hard rock of her childhood.

White Island lighthouse and Laighton cottage.

276

It was a good move for everyone, perhaps most of all for her father, Thomas, who, with his brother Joseph, had bought the islands of Hog, Smutty Nose, and Malaga. Defeated in a political campaign, Thomas Laighton then moved to the most westerly of the Isles of Shoals—White Island—and there he became the lighthouse keeper. He could, like a feudal lord, overlook his own islands and try his hand at fishing.

The first winter was a lonely one; a storm wrecked the hen houses that had been patiently erected, rowboats were washed away and even the family cow had to be brought into the kitchen to protect her from the weather. The family awaited the signs of spring each year: the first glimpse of poor man's weatherglass—we would call it scarlet pimpernel now—morning glories, wild roses, bluets, elderbloom, and finally goldenrod. The flowering of each rock flower meant more than the rich gardens and the little village that had been left behind. And the gulls; the gulls became her dearest friends for a while. She would watch them circle and cavort in the sky, swoop down with a sudden sharp beauty, and then take off again. Occasionally on the little beach she would see a sandpiper "skimming along with a sweet mournful cry," she said. She would listen to the gulls hailing her with a cry at night while the waves around the rock would whisper "wild and sweet."

White Island light and Sevey's Island viewed from Appledore.

277

White Island was one of the typical lighthouse islands of the New England coast. Thomas Laighton manned it from 1839 to 1841. The rocks of these shoals were dangerous and the isolation of the Laightons was almost complete. They would wait eagerly for the government inspector who brought his supplies and a touch of the outside world.

Occasionally, too, a schooner carrying the oil that would fuel the light would make its way laboriously to the rock. The child, Celia, used to love to watch the oil being stored in the cellar below the tower. She learned how the oil lamps had to be filled and the wicks trimmed. She would clean the windows assiduously so that the light that sprang from them could be seen far and wide. She learned early the law that had been enacted in 1821: "If any person or persons shall hold out or show any false light or lights, or extinguish any true light with the intention to bring any ship or vessel, boat or raft, being or sailing upon the sea, into danger or distress of shipwreck, every such person so offending shall be deemed guilty of felony and shall on conviction thereof punished by a fine not exceeding four-thousand dollars and imprisonment and confinement to hard labor not exceeding ten years according to the aggravation of the offense." In those days there were still rumors of evil lighthouse keepers, particularly those placing false lights; and in the Bahamas it was rumored that they were still luring ships off their course and onto reefs, using a method of tying a lantern to a horse's tail and walking the animal around in a circle, so dooming many a ship.

The United States had been one of the last to develop a decent lighthouse service. By the time the Laightons took over their island, there had been many improvements; reflectors and lamps had been introduced and engineering skills had come into play. Celia could not actually man the light; there was a law that "women and servants must not be employed in the management of the lights except by the special authority of the department." But she learned its workings well. She was a good observer and there was a great deal to see, not just the spring flowers and the tempestuous curling of the waves, but the ways of the island. Cattle had to be watched carefully, for example. Attracted to a light at night, the animals would so destroy the rooted grass which held down the sand dunes that the beacon was easily exposed to destruction. Every lighthouse keeper knew, too, that ghastly thump as seafowl dashed against the lantern glass, blinded by the lights. If Celia could not actually work the light, she was gloriously well informed about it. She watched her father as he made the daily entries in the journal which was required of every keeper, and she knew every direction for cleaning, placing, and removing the lamp chimneys that kept the lantern free from ice and snow, and the fundamental duties of the lightkeeper, which read:

278

The lighthouse and light vessel lamp shall be lighted and the lights exhibited for the benefit of mariners punctually at sunset daily. Lighthouse and light vessel lights are to be kept burning brightly free from smoke and at the greatest obtainable heights during each entire night from sunset to sunrise. . . .

The height of the flame must be frequently measured during each watch at night by the scale graduated in inches and tenths of an inch with which keepers are provided. . . .

All lighthouse and light vessel lights shall be extinguished promptly at sunrise and everything put in order for the lighting in the evening by ten o'clock a. m. daily.

In 1841 the family moved to Smutty Nose Island, a far more hospitable island with an old inn that the Laightons took over. By the time Celia was twelve they had moved once again—to Appledore Island, richer in vegetation and stretching a mile from end to end. Pleased with the idea of innkeeping, her father now built a new hotel and on June 15, 1848, Appledore House was opened to the public to become one of the major resort hotels of the nineteenth century. There was a new man on the island, Levi Thaxter, and Celia was to marry him, but not, however,

Lighthouse at East Quoddy, Campobello, New Brunswick.

until Thaxter and Laighton made an agreement about the property on Appledore. Thaxter, who had tutored Celia in her schoolwork, soon became strongly attracted to his student and although she was not yet sixteen, they were married on a beautiful island fall day in 1851. For the rest of her life, between raising a family and living first on the mainland and then on the island, her thoughts would turn to writing and specifically writing about the Isles of Shoals, that she had known so well. She soon made this small group of islands, just nine miles from the Portsmouth coast, well known throughout the world. That summer of 1852 the islands were visited, as they would continue to be, by many famous people including Nathaniel Hawthorne, who left us a record of his visit:

Leaving Portsmouth at 10:30 A.M. we sailed out with other passengers to the Isles of Shoals. The wind not being favorable, we had to make several tacks before reaching the islands, where we arrived about two o'clock. We landed at Appledore, on which is Laighton's Hotel—a large building with a piazza or promenade before it, about an hundred and twenty feet in length, or more, yes, it must be more. It is an edifice with a center and two wings, the central part upwards of seventy feet. At one end of the promenade is a

280

The Beginning of the Jaunt, or "I smell you in the dark." Samuel Johnson and James Boswell walking up the High Street of Edinburgh on August 14, 1773. (From an engraving by Thomas Rowlandson)

covered verandah, thirty or forty feet square, so situated that the breeze draws across it from the sea on one side of the island to the sea on the other, and it is the breeziest and [most] comfortable place in the world on a hot day. There are two swings beneath it, and here one may sit or walk, and enjoy life, while all other mortals are suffering. . . .

After dinner, which was good and abundant, though somewhat rude in its style, I was introduced by Mr. Laighton to Mr. Thaxter, his son-in-law, and Mr. Weiss, a clergyman of New Bedford, who is staying here for his health. They showed me some of the remarkable features of the island, such as a deep chasm in the cliffs of the shore, towards the southwest; also a monument of rude stones, on the highest point of the island, said to have been erected by Captain John Smith before the settlement of Plymouth. . . .

This island (Appledore) is said to be haunted by a specter called "Old Babb." He was one of Captain Kidd's men, and was slain for the protection of the treasure. Mr. Laighton said that, before he built his house, nothing would have induced the inhabitants of another island to come to this after nightfall. The ghost especially haunts the space between the hotel and the cove in front. There was, in times past, great search for the treasure, which continues.

Boulder Beach, one of Walt Whitman's Long Island beaches. (Photograph by R. Paterline)

Mr. Thaxter had once a man living with him who had seen "Old Babb," the ghost. He met him between the hotel and the sea, and describes him as dressed in a sort of frock, and with a very dreadful countenance, who always carried a butcher knife.

Last evening we (Mr., Mrs., and Miss Thaxter) sat and talked of ghosts, and kindred subjects; and they told me of the appearance of a little old woman in a striped gown, that had come into that house a few months ago. She was seen by nobody but an Irish nurse, who spoke to her, but received no answer. The little woman drew her chair up toward the fire, and stretched out her feet to warm them. By and by the nurse, who suspected nothing of her ghostly character, went to get a pail of water; and, when she came back, the little woman was not there. It being known precisely how many and what people were on the island, and that no such little woman was among them, the fact of her being a ghost is incontestable.

Very few accidents happen to the boats or men,—none, I think since Mr. Thaxter has been here. They are not an enterprising set of people, never liking to make long voyages. Sometimes one of them will ship on a voyage to the West Indies, but generally only on coastwise trips, or fishing or mackerel voyages. They have a very strong local attachment, and return to die.

Thatcher's Island (Cape Ann, Massachusetts) light and fog signals.

They are now generally temperate, formerly very much the contrary. . . .

Mrs. Thaxter tells me that there are several burialplaces on this island; but nobody has been buried here since the Revolution. Her own marriage was the first one since that epoch, and her little Karl, now three months old, the first-born child in all those eighty years.

The best recorder of island living was Celia Thaxter herself. She was fascinated, as all of us are, by the peculiar local pronunciations you will often get in an island colony and particularly in those New England colonies that have remained so isolated and alone. The Shoalers in particular had a pronunciation both peculiar and witty. Perhaps like the Irish, who felt cut off from the outside world so that they perfected and enriched their use of words, they expended their language in a glorious excess. They cultivated it; their speech had many forms to express all variations of emotion, pain, anger, or amusement. The "unhesitating profanity," as Celia Thaxter called it, was sometimes chilling to the vein—to the nineteenth-century vein, that is. But even she had to be delighted by its power to amuse intensely. In the first place the intonation was completely different. It was not a Yankee drawl, nor was it sailor talk. The commonest word had a different twist. The birds, for example: A swallow would fly over, but it was not a swallow, it was a "swallick"; a sparrow became a "sparrick," and the great tempests that roared over the island were called "tan toasters" (the derivation lost in mist). Common names often added a suffix to every name: Benny was "Bennaye," Billy, "Billaye."

Gatherers of driftwood, Grande Isle, Louisiana. (Nineteenth-century print from the authors' collection)

283

Some ten miles off the coast of Portsmouth, New Hampshire, are the famed Isles of Shoals, of which Star Island is the most prominent. There are rumors pirate treasure is secreted somewhere within their rugged, boulder-strewn surroundings. (State of New Hampshire photograph by Eric M. Sanford)

A view of the beach on Inisheer, Aran Islands, County Galway. (Courtesy Irish Tourist Board)

284

Aran woman with her grandsons on Inisheer, Aran Islands, Ireland. (Courtesy Irish Tourist Board)

"The Gulf Stream" by American artist Winslow Homer (1899). (Courtesy Metropolitan Museum of Art, New York City)

Nicknames were even more common. No one was completely protected from their observations and irreverence. When the minister of, as Celia Thaxter said, "Methodist persuasion" lived among them, his wife, tall and thin, was immediately christened "Legs"; she was never called anything else. Legs was actually pronounced "Laigs" and the calendar of the year would be determined by the time that "Legs has gone to Portsmouth." Or social activity by "Legs has got a new gown." Such nicknames were used not only in private conversation, but directly to the lucky soul who had inspired them. There was, for example, one man on the island, one of the greatest braggarts of all times according to the Shoalers, who christened him simply "Brag." He was never known by any other name and his wife became "Mrs. Brag." In a few years Brag and Mrs. Brag began to call themselves by that name; their given name was washed out into the sea of time and the new one used by them with great affection.

The old name of Grandsire hung on for a long time on the island, the grandfather being called "Grans" ("Tell your Grans his dinner is ready") and the grandmother, "Gwammaye."

Celia Thaxter used to go through the islands rather as the Irish did through the Gaeltacht in the nineteenth century, collecting some of the lore that she most loved. An old woman told her that her home was put together skimpily. She said, "Lor't warn't never built. Twas only hove together."

285

ISLANDS

Being creatures of a marine world, the old Shoalers walked with a singular gait. They had that curious crooked stability that comes from standing in small boats and an almost contorted lumbering that came from climbing the rocks. Even the oldest of men could leap wildly from rock to rock and in the few times they went to the mainland of Portsmouth for supplies they were always identified by the way they walked, not on the roadway itself, but jumping up and down as children do on the rocks that decorated the lawns by the pavement. They were all round-shouldered from rowing and from bending over fish tables, cleaning, splitting, salting, and packing the fish that they had caught so laboriously. They were, as are all islanders, devoted to their boats. "Oh she's a pretty piece of wood" was the greatest praise a man could bestow on his boat. Another would describe his smooth-sailing dory by saying, "She strikes the sea and comes down like a pillow."

The fish haul was called in those days "a fare of fish" and hake-fishing in small dories was the main livelihood of the islanders. Their sustenance was fish, milk from the few cows, and eggs. Celia Thaxter used to love to go into a household and hear some old grandmother call up to her husband in the loft, which served as sleeping quarters and a nesting haven for the hens, "Come, Benaye, fetch me down them heens' aigs"; and the answer would come, "I can't find no aigs. I've looked e'en in the bed and e'en under the bed and I can't find no aigs."

Even in child rearing the local expressions were vivid. Celia Thaxter, when she became famous, was once asked to suggest a highfalutin name to be given to a firstborn boy. She could choose no better than the name of the then current President and used to rejoice in hearing the boy summoned by his mother with the words: "Taylor, if you don't come into the house this minute, I'll slat your house off." This tender expression, as Celia Thaxter tells us, was used commonly by the fishermen who slatted delicate-gilled fish off the hook.

One islander who tried to give a native of Rye an opinion on some matter of moment was answered with withering scorn, "You don't know nothin' about it. What do you know? You never seen an apple tree all blowed out." There were neither frogs nor toads at the Shoals and an islander walking in the streets of Portsmouth suddenly saw a toad. "Mr. Berraye, what kind of bug do you call that? Damned if I ever did see such a bug as that, Mr. Berraye."

Later when folklorists visited obscure islands in the world, they often saw the folk music or the local speech with a far more romantic eye than Celia Thaxter was ever to do. In those days there was music on the island, but there was no sentiment about the performers in Celia's mind. They were, to put it bluntly, quite dreadful.

Many of the old gentlemen considered themselves splendid musicians, but perhaps the best description of then was that they were "interminable singers." When they put their violins under their chins Celia Thaxter admitted quite frankly that they had no more capability of playing them than they might an old codfish. They all sang with the tune pitched high and then all complained, "Tew high, Bill, tew high."

"Woll, *you* strike it, Obid," Bill would answer.

Obid would strike exactly the same note and Bill would reply, with great graciousness, "Damned if he ain't got it."

Celia Thaxter said, "One could but wonder whence these queer tunes came—how they were created; some of them reminded one of the creaking and groaning of the windlasses and masts, the rattling of rowlocks, the whistling of the wind among cordage."

North of the Shoals is a land of poetry, of pine trees and rock. The names of the islands themselves imply history, and in their glorious nomenclature you will find excitement of yesterday and the pleasure of now. You will find one called Pound of Tea, another a Junk of Port; others are Stepping Stones, Old Dick, Little John, White Bull, Brown Cow, Whaleboat, Stave and Clapboard. There are thousands of Maine islands "hoved up," as they say there, by the sea.

One of the almost perfect guides to the area is Sarah Orne Jewett, who captured Maine so serenely in *The Country of the Pointed Firs.* Although she was a mainlander, she knew the islands well and because she is one of the greatest regional writers we have ever had she is an enchanting companion amid the pines of Maine. Who can ever forget her characters who inhabited strange islands—so often wild, desolate characters, more lonely than a lone pine tree, more rugged than the rock of Monhegan? Take Joanna, for example, of Shell Heap Island. Whenever you ask about some of the small islands on the coast of Maine you get the kind of information that Sarah Jewett knew so well. Where is Shell Heap Island? you will ask. It is, of course, almost imaginary—it could be anywhere along the Maine coast—but the specific directions are "Somewhere about three miles from Green Island, right off shore. You would never know it was there, dear; it is off the thoroughfares in a very bad place to land at best."

On some of these islands—and Shell Heap was one of them—there would be strange hermits; some who had been crossed in love and some just doomed to melancholy. A few hermits just wanted to get away from folks; a few thought they weren't fit to live with anybody; others simply wanted to be free. Miss Jewett's Joanna lived on such an island. She had been crossed in love. Joanna took herself to an island of thirty acres, rocks and all, with nothing but the salt spray to succor her, nothing but a mud

287

flat from which to dig a good clam occasionally; and nothing for company but an old sheep or two and a bad edge of weather.

Joanna had simply taken an old boat of her father's, "lo'ded in a few things and off she put all alone, with a good land breeze, right out to sea." Her father ran down to the beach and stood there crying like a boy at seeing her go, but she was out of hearing. She'd never set foot on the mainland again as long as she lived.

Joanna had another suitor, who tried occasionally to make some effort to fetch her from the island. He would sail close to it, and sometimes he would throw on shore pretty things that a woman would like. He could see Joanna come down from a rock and walk around, ignoring what he had tossed on the beach. The next morning while he was out makerel fishing, he would find that his present was gone. In the spring he did better; he got her some hens and chickens, but he could never entice her off her rock. They say she was a girl who had always had friends, but her only friends became the few hens she had. She even turned against the sheep. There was not enough grass on the island, and she did not like to see them suffer. There were some who called her a fool; "Imagine making friends of hens," they would say.

Once the minister went out to try to help her. Elmira Todd rowed him out, but he was the landlubber type—one gust of wind frightened him and he screeched for help. Joanna was decent enough to him when he came, but she had a blank eye when he asked her whether she thought it was Christian to stay on the island, and he went off defeated. So she lived there the rest of her life thinking that she should return to the mainland when her time came to die, but it was not to be—she died and was buried on the island.

That was Shell Heap Island, and then there was Miss Jewett's Green Island, one of those beautiful islands that you should visit when the winds are light northeast, the kind of breeze that will take you straight out, turn around by afternoon to the southwest, and fetch you home pretty late in the day. Where are those islands? Just somewhere in the glorious imagination of American writers where the tide is setting in, where the gulls have the freedom of the wind, and the world is as fresh as early grass.

But the Scottish islands are very real indeed. The trip to those islands was to be one of the high points of James Boswell's life. He was to go with a good companion, a very good companion, and the journey had been planned for a long time. Perhaps it was planned on that very memorable day of Monday, May 16, 1763, when James Boswell first met the great sage Dr. Samuel Johnson. If it was planned, it was planned only in the inventive mind of Boswell himself. The thought of going to

Scotland surely had never occurred to Dr. Samuel Johnson. Quite the contrary, he was extremely negative about Scotland, as could be told from the very first record we have of his meeting with Boswell. It is recorded by another protégé of Johnson's, Arthur Murphy.

Upon another occasion this writer went with him (Johnson) into the shop of Davies the bookseller in Russell Street, Covent Garden. Davies came running to him almost out of breath with joy. "The Scots gentleman is come, sir; his principal wish is to see you; he is now in the back parlour." "Well, well I'll see the gentleman," said Johnson. He walked towards the room. Mr. Boswell was the person. This writer followed with no small curiosity.

"I find," said Mr. Boswell, "that I am come to London at a bad time when great popular prejudice has gone forth against us North Britons. But when I am talking to you, I am talking to a large and liberal mind and you know I cannot help coming from Scotland."

"Sir," said Johnson, "no more can the rest of your countrymen."

But James Boswell was ingratiating and obstinate. He would have Johnson for a friend and he would, as he planned, take Johnson to the Scottish islands.

The task seemed formidable. The good doctor enjoyed "the pleasures of lying abed daily until noon." All forms of physical activity were a horror to him. London was the only place to exist. Scotland and particularly the Hebrides were filled with strange wild places, the ground was unstable under foot, the seas were menacing with whirlpools. Yet Boswell convinced him. It took ten years, however. The friendship or discipleship between Johnson and Boswell had grown richly. Boswell was married to a respectable Scottish woman who scoffed at Boswell's adulation: "I have," she said, "seen many a bear led by a man. I have never before seen a man led by a bear."

And indeed in those days Johnson did look like an overgrown bear. He was sixty-five, the friend of many and the enemy of perhaps even more. His appearance, even for those days, was extraordinary:

He wore a full suit of plain brown clothes, a large, bushy grayish wig, a plain shirt, black worsted stockings and silver buckles. Upon this tour when journeying he wore boots and a very wide brown cloth greatcoat with pockets which might have held the two volumes of his folio dictionary and he carried in his hand a large English oak stick.

Johnson had planned to bring on the trip two pistols with the appropriate bullets and gunpowder to take care of the aboriginals that they would run into in the islands. Starting their journey on August 18, a very inauspicious time for Scotland and the Scottish islands, they re-

Visitors take time to photograph the Annaberg Ruins in the Virgin Islands. (National Park Service photograph by M. W. Williams)

turned to Edinburgh on Tuesday, November 9. "I am more an adventurer than you, sir," said Johnson. And certainly as we start them on their voyage, we can almost hear the heaving, corpulent misery as Johnson puts himself aboard the small boat. Certainly at the start of the voyage Johnson had what Boswell called a strange tranquillity. When Boswell queried him he said, "Sir, when a man retires into an island he is to turn his thoughts entirely on another world. He has done with this."

For Johnson there was a kind of terrible finality about his surrender to the islands. He was putting himself in the hands of aboriginals. For his part Boswell was putting himself in the hands of fellow countrymen and very good hosts. Almost immediately he drank too much. By the time noon cleared, he would say regularly throughout his diary, he would grow better.

The lairds and their ladies had a rather difficult time understanding this bear in his capacious coat. He was as outspoken as they, but without, sometimes, that delicate use of the point of a dirk instead of the verbal dagger that was Johnson's weapon. On Skye they were most concerned about Dr. Johnson. His illness had surely come from the fact that he had not worn a nightcap, so he was made a large flannel one. He was urged, too, in the usual Scottish fashion to take a wee dram, and Lady MacLeod said: "I am sure, sir, you will not carry it too far."

Johnson replied: "Nay, madame, it carried me."

And he would touch hardly anything on the trip, with the good Boswell making up for it during the entire time. The chill of the islands had reached Johnson's very bones, and he would often go out and gather his own peat.

As Johnson traveled he observed. He realized that the system in the Highlands was wrong. The Highland chiefs raised rents, but there had been no real effort to see that the people were properly employed.

Some of the small islands attracted Johnson strongly. On one he contemplated, at least for a brief moment, setting up an Episcopal school with Boswell. One small island was offered by the MacLeods to Johnson:

Boswell remembered he talked a great deal of this island, how he would build a house, how he would fortify it, how he would have cannon, how he would plant, how he would sally out and take the Isle of Muck.

But then Johnson would rebel against island life. He repeated to Boswell: "Do you remember a song which begins, 'Every island is a prison strongly guarded by the sea. Kings and Princes for that reason prisoners are as well as we.'?"

And indeed the feudal islands of that period were almost prisons—prisons of poverty for many, sometimes of loneliness for the aristocratic laird. Johnson, too, seemed well aware of the absentee landlords of the islands.

"I told Mr. Johnson," said Boswell, "that Sir Alexander (master of Skye) said to me once that he left Skye with the blessings of his people."

Johnson simply replied somewhat bitterly, "You'll observe this is when he left it. It is only the back of him that they bless."

But when the laughter subsided, Johnson moaned: "I want to be on the mainland and go on with existence; this is a waste of life."

Occasionally they would meet someone who had never been on the mainland; Johnson was shocked. Boswell reminded him, however, that he, too, had never left his native island. Johnson argued for London: "By seeing London I have seen as much of life as the world can show."

"You have not seen Peking," quarreled Boswell.

"Sir," said Johnson in a sort of anger, "what is Peking? Ten thousand Londoners would drive all the people of Peking, they would drive them like deer."

Finally there was a good day for passage to Mull, but as they boarded the boat the waters grew dangerous again. But they arrived safe in the harbor of Tobermory. As with many journeyers through the islands, Boswell's spirits recovered once he was on land again. Johnson said: "Boswell is now all alive. He is like Antaeus, he gets a new strength whenever he touches land."

As they sailed, one day there was a black barren rock in their path. Johnson called to Boswell and said, "This shall be your island and it shall be called Inchboswell." Then he laughted with a strange appearance of triumph.

For his part, Johnson was getting crankier. When one of the lads bragged that Scotland had the advantage over England of having more water, he simply replied: "Sir, we would not have your water to take the vile bogs which produce it. You have too much. A man who is drowned has more water than either of us."

They arrived eventually on the mainland and Johnson was so relieved he said, "Come, let me know what makes a Scotsman happy," and drank a whiskey with Boswell. The trip to the Hebrides was over. Johnson put up a leg on each side of the grate and said loudly enough for Boswell to hear: "Here am I, an Englishman, sitting by a cold fire."

The days of the rain and mist and storms and peat of the Hebrides were over and yet the fantasies about them would pursue British and American travelers for the rest of their lives.

292

There is a small group of islands off the coast of Ireland, across Galway Bay, that turned a talented man into a writer with a genius for words. His name was John Millington Synge. He had probably heard first about the wastes of Aran from his uncle, the Reverend Alexander Synge, who had been a Protestant minister in those stone-carved islands that were to attract not only Synge but many writers of Ireland and the world.

It was in the year 1898 that he first stepped on the shore of the Aran Islands. He wrote a friend, "Did you ever hear tell of a place across the bit of the sea where there is an island and the grave of the four beautiful saints?"

In later years people were to say about him, "What did he really know about the islanders? Hadn't he spent most of his time resting alone in the sun, and in the evenings—those curious island evenings in which the Arans seemed so close to the sky—didn't he simply mope under the stars?" He did a great deal of lying around in the sun whenever there was sun, which was rare enough, and he did mope about under the stars. One day he said:

After sunset the clouds broke and the storm turned to a hurricane. Bars of purple clouds stretched across the sand where immense waves were rolling from the west. There was a bay full of green delirium and the Twelve Pins were touched with mauve and scarlet in the east.

The suggestion from this world of an articulate power was immense, and now at midnight when a wind is abating, I am still trembling and flushed with exultation. . . If anything serious should happen to me, I might die here, and be nailed in my box and shoved down into a wet crevice in the graveyard before anyone could know it on the mainland.

In the morning it would clear again and he would see the women in their red dresses. One adolescent girl who sat in a doorway always caught his eye as the sunlight fell on the net she was mending. They used to say that there was a connection between the strange wild beauty of the women of Aran and the wild mythology of the place.

These people were completely dependent upon the weather. Bad weather was a threat to their very existence. Men could drown in a flash, families could be decimated, all of the men of a small configuration of houses annihilated by one fierce wave. The waves educated them. "This continual danger which can only be escaped by extraordinary personal dexterity has had considerable influence on the local character, as the waves have made it impossible for clumsy, foolhardy or timid men to live on these islands." _293

We first went to Aran carrying with us a little green pocket edition of Synge's works. It was a highly appropriate introduction. For a long time we used to stay on the shingled beach of Galway Bay, that magnificent bay where Grace O'Malley, the famous lady pirate, had autonomy.

The ferry to Aran was still irregular; it did not run in bad weather. And there were few to catch it in any case. World War II was over, but barely. No tourist associations attracted people with multicolored advertisements of the islands of the world. One had to be attracted almost out of an innate need, or by the first and still the most commanding travel guide to any spot, the words of a writer who had loved it dearly.

For a while the people of Aran had considered themselves part of the great outside world. The outside world had come close to them with cameras and the mechanics of twentieth-century living. It was as though, if only for a moment, these enormous rocks had been deposited closer to the civilized world and were as they might have been in geological years gone by—a part of Ireland's mainland. But during the war the islands had been cut off again, and when we arrived there on that August day there was no one to meet the ferry except a few islanders shuffling around, one still wearing the pampooties, the hide shoe which is so familiar or was on the islands in those days.

If the island women's faces have that strange, rather withdrawn look of a madonna, the men's faces seem to have a rough, hurt look. Even their clothes, when one was close enough to feel the roughness of them, pressed as we were on a very small boat, had the smell of seaweed and burnt turf, imported from Connemara, that was still used in the hearths.

We remembered Synge's words when he said on his first trip to the islands, "It gave me a moment of exquisite satisfaction to find myself moving away from civilizations in this rude canvas canoe of a model which has served primitive races since men first went on the sea."

Sometimes a writer seeks out a place; sometimes he seeks out his own voice. Synge found a place and a voice—the intonation of his own culture that was to burst out into not only the flowering of his talent, but into that great Irish renaissance of writing that the very island of Ireland inspired.

Ever since the ninth century Ireland had been a lonely voice when the rest of Western civilization was speechless in darkness. She had been the great heiress of the classical and theological inspiration of the past. Life, said an early Irish poem, held only three candles, but those candles illumined every darkness. They were truth, knowledge, and nature.

The landscape of Ireland has determined its history. "Ireland lieth aloof in the West Ocean" wrote an early historian. The island is the last body of land on the continuous continental shelf of Europe. At times this isolation protected her; it preserved a culture when the rest of Europe was in chaos, but like all islands she was subject to invasion. The invaders left their marks on the landscape: Danish, Norman, English place names and towns cover the countryside.

The invaders were interested only in those sections of the country where coastal towns might be established, or where the land was fertile for crops. They were not interested in the great limestone belt which stretches across Ireland, a relic of the ice age, which still supports only mile after mile of peat bog. A poor source of fuel, but none the less a source, this stretch of land is an umbilical cord which ties the people only to poverty. Nor were they interested in those areas which today are known as the Gaeltacht, the Irish-speaking districts where the rocks defy cultivation. But it was in just those areas that the native Irish sought refuge.

Daniel Corkery described this part of Ireland as the Hidden Ireland, and it was from these sections that another stream of great Irish poetry developed. Like the early lyric poetry, these later poems are in Gaelic, and their simple subject matter is that of the countryside. However, in a good 500 years only a few names stand out. Raferty, who has been anthologized in translation, wandered, blind as Homer, the length and breadth of the country. As he went he recited his poems. "I am Raferty the Poet," he said, "full of hope and love." There were Owen of the Sweet Mouth and Brian Merriman. But principally there was the body of literature that was anonymous, passed on generation after generation by the village shanachie, or storyteller.

Sometimes there were fairy stories. Ireland has a quality of enchantment; the mists hang low, bathing the mountains, breathing up from the rivers. In a country in which one can walk literally through clouds, it is simple to see how that landscape has inspired fairy stories.

The tradition of the Irish storyteller was a noble one. It descended in one straight line from the ancient "fildh," the poets, priests, and diviners whose bardic order in ancient Ireland was indispensable to the great kings. The bards once roamed from castle to castle, as the shanachie later did from hut to hut, reciting the great hero stories of Finn and Ossian, the love stories of Deirdre of the Sorrows. Many different versions of these stories existed, the locales and the adventures of the participants changing to make the stories applicable to different counties. Ireland even in recent years was the last country in Western Europe where stories and poems might be collected whose ancestry stretched back a good 1100 years. Today the Irish Folklore Commission is busy recording the last remnants of this tradition, using as their guide until recently a large electrical map of Ireland. Where electricity had made inroads, the storyteller had deserted his post for the knob of the radio.

295

Change came slowly to Ireland. Except for the Gaelic poets of the seventeenth and eighteenth centuries and the storytellers, literature disappeared. The Anglo-Irish writers wrote *about* rather than *of* Ireland, and to them the land and its people were barbarous, or if not barbarous, vastly entertaining. The Irish, perverse in their unhappiness and deprivation, contributed in their own way to this picture of the "stage Irish."

But Ireland as it sought freedom also sought identity, and the result in literature combined the best features of her entire tradition. The period known as the Irish Renaissance contained many abortive forays backward into the earlier cultural period of Ireland, but it succeeded majestically in synthesizing the best of the past with the new Ireland. Writers and poets became alert to their own heritage, aware of their country's history in tradition, and from the synthesis of time, place, and a gift for words made up for a long period of enforced silence.

Now they wrote in English and what they wrote was a major contribution to the literature of the entire English-speaking world. The greatest poet among them was William Butler Yeats. Yeats was a

296   Petroglyphs by a spring near Reef Bay in the Virgin Islands. (National Park Service photograph by Jack E. Boucher)

countryman. The countryside fed him, nursed his talent, rewarded his observations. His early poems, romantic and beautiful, harked back to the poems of the ancient Irish hermits. He mourned for a "lowly hut" where the swans of Coole might visit him as the robins and blackbirds had entertained his ancestors. As his talent developed, his poems took on a strength and sharpness which assured him his place as the greatest poet of his time. Yeats's love for the Irish landscape persuaded him to suggest to the expatriate J. M. Synge that the latter desert Paris for the western islands of Ireland. From his life in Connaught and Wicklow, from his stay on the Aran Islands, Synge, perhaps more than any other writer, collaborated with nature; all art, he said, was such a collaboration—with nature and with the people of a place.

It was Lafcadio Hearn, having spent his childhood in Ireland, who gave a voice to a wide variety of islands—from Martinique to Japan.

He was a lover of islands. In addition to having an Irish heritage he learned perhaps a further love of islands from his mother, who was Greek. As a young man he moved to Cincinnati, where he began to write a series of brilliant essays on the raw years of Cincinnati after the Civil War. But he longed for something else. Deep within him he always

The beach at Trunk Bay in the Virgin Islands. (National Park Service photograph by Jack E. Boucher)

seemed to long for the peculiar call of an island. Then suddenly he found it. He went to New Orleans and then had his first summer at the once superb island in Louisiana called Grande Isle.

He was to write about many islands—for example, the British West Indies, about which he said, "All this majesty of light and form and color will surely endure." He was simply and completely in love with what he called "the magic of words." The magic of words—and the beauty of places. He would write endlessly about the home in which he was to die—Japan, where he married a Japanese woman. He was to sing of all the islands of the West Indies, but he was never to be more struck, more taken by the simple beauty of water, than he was by the Gulf of Mexico. "There is something unutterable in this bright Gulf air."

Traveling south from New Orleans to the islands, you pass through a strange land into a strange sea, by various winding waterways. You can journey to the Gulf by lugger if you please; but the trip may be made much more rapidly and agreeably on some one of those light, narrow steamers, built especially for bayou-travel, which usually receive passengers at a point not for from the foot of old Saint Louis Street, hard by the sugar-landing, where there is ever a pushing and flocking of steam-craft—all striving for place to rest their white breasts against the levee, side by side—like great weary swans. But the miniature steamboat on which you engage passage to the Gulf never lingers long in the Mississippi; she crosses the river, slips into some canal mouth, labors along the artificial channel awhile, and then leaves it with a scream of joy, to puff her free way down many a league of heavily shadowed bayou. Perhaps thereafter she may bear you through the immense silence of drenched rice-fields, where the yellow-green level is broken at long intervals by the black silhouette of some irrigating machine; but, whichever of the five different routes be pursued, you will find yourself more than once floating through sombre mazes of swamp-forest—past assemblages of cypresses all hoary with the parasitic tillandsia, and grotesque as gatherings of fetich-gods. Ever from river or from lakelet the steamer glides again into canal or bayou—from bayou or canal once more into lake or bay; and sometimes the swamp-forest visibly thins away from these shores into wastes of reedy morass where, even of breathless nights, the quaggy soil trembles to a sound like thunder of breakers on a coast: the storm-roar of billions of reptile voices chanting in cadence—rhythmically surging in stupendous crescendo and diminuendo—a monstrous and appalling chorus of frogs! . . .

Panting, screaming, scraping her bottom over the sand-bars—all day the little steamer strives to reach the grand blaze of blue open water below the marsh-lands; and perhaps she may be fortunate enough to enter the Gulf about the time of sunset. . . .

298

Beyond the sea-marshes a curious Archipelago lies. If you travel by steamer to the sea-islands to-day, you are tolerably certain to enter the Gulf by Grande Pass—skirting Grande Terre, the most familiar island of all, not so much because of its proximity as because of its great crumbling fort and its graceful pharos: the stationary White-Light of Barataria. Otherwise the place is bleakly uninteresting: a wilderness of wind-swept grasses and sinewy weeds waving away from a thin beach ever speckled with drift and decaying things—worm-riddled timbers, dead porpoises. Eastward the russet level is broken by the columnar silhouette of the lighthouse, and again, beyond it, by some puny scrub timber, above which rises the angular ruddy mass of the old brick fort, whose ditches swarm with crabs, and whose sluice-ways are half choked by obsolete cannon-shot, now thickly covered with incrustation of oyster shells. . . . Around all the gray circling of a shark-haunted sea.

Southwest, across the pass, gleams beautiful Grande Isle: primitively a wilderness of palmetto (latanier):—then drained, diked, and cultivated by Spanish sugar-planters; and now familiar chiefly as a bathing-resort. Since the war the ocean reclaimed its own;—the cane-fields have degenerated into sandy plains, over which tramways wind to the smooth beach—the plantation-residences have been converted into rustic hotels, and the Negro-quarters remodeled into villages of cozy cottages for the reception of guests. But with its imposing groves of oak, its golden wealth of orange-trees, its odorous lanes of oleander, its broad grazing-meadows yellow-starred with wild camomile, Grande Isle remains the prettiest island of the Gulf; and its loveliness

The palace of Diego Colon, the son of Christopher Columbus. Its location is in
Santo Domingo in the Dominican Republic. (Courtesy Grace Line)

"Bermuda Sloop" by American artist Winslow Homer (1836–1910). (Courtesy
Metropolitan Museum of Art, New York City)

is exceptional. For the bleakness of Grande Terre is reiterated by most of the other islands—Caillou, Cassetete, Calumet, Wine Island, the twin Timbaliers, Gull Island, and the many islets haunted by the gray pelican—all of which are little more than sand-bars covered with wiry grasses, prairie-cane, and scrub timber. Last Island (L'Ile Dernière)—well worthy a long visit in other years, in spite of its remoteness—is now a ghastly desolation twenty-five miles long. Lying nearly forty miles west of Grande Isle, it was nevertheless far more populated a generation ago: it was not only the most celebrated island of the group, but also the most fashionable watering-place of the aristocratic South;—today it is visited by fishermen only, at long intervals. Its admirable beach in many respects resembled that of Grande Isle to-day; the accommodations also were much similar, although finer: a charming village of cottages facing the Gulf near the western end. The hotel itself was a massive two-story construction of timber, containing many apartments, together with a large dining-room and dancing-hall. In the rear of the hotel was a bayou, where passengers landed—"Village Bayou" it is still called by seamen;—but the deep channel which now cuts the island in two a little eastwardly did not exist while the village remained. The sea tore it out in one night—the same night when trees, fields, dwellings, all vanished into the Gulf, leaving no vestige of former human habitation except a few of those strong brick props and foundations upon which the frame houses and cisterns had been raised. One living creature was found there after the cataclysm—a cow! But how that solitary cow survived the fury of a storm-flood that actually rent the island in twain has ever remained a mystery. . . .

300

On the gulf side of these Islands you may observe that the trees—when there are any trees—all bend away from the sea; and, even of bright hot days when the wind sleeps, there is something grotesquely pathetic in their look of agonized terror. A group of oaks at Grande Isle I remember as especially suggestive: five stooping silhouettes in line against the horizon, like fleeing women with streaming garments and wind-blown hair—bowing grievously and thrusting out arms desperately northward as to save themselves from falling. And they are being pursued indeed;—for the sea is devouring the land. Many and many a mile of ground has yielded to the tireless charging of Ocean's cavalry; far out you can see, through a good glass, the porpoises at play where of old the sugar-cane shook out its million bannerets; and shark-fins now seam deep water above a site where pigeons used to coo. Men build dikes; but the besieging tides bring up their battering-rams—whole forests of drift—huge trunks of water-oak and weighty cypress. Forever the yellow Mississippi strives to build; forever the sea struggles to destroy—and amid their eternal strife the islands and the promontories change shape, more slowly, but not less frantically, than the clouds of heaven.

And worthy of study are those wan battle-grounds where the woods made their last brave stand against the irresistible invasion—usually at some long point of sea-marsh, widely fringed with billowing sand. Just where the waves curl beyond such a point you may discern a multitude of blackened, snaggy shapes protruding above the water—some high enough to resemble ruined chimneys, others bearing a startling likeness to enormous skeleton-feet and skeleton-hands—with crustaceous white growths clinging to them here and there like remnants of integument. These are bodies and limbs of drowned oaks—so long drowned that the shell-scurf is inch-thick upon parts of them. Farther in upon the beach immense trunks lie overthrown. Some look like vast broken columns; some suggest colossal torsos embedded, and seem to reach out mutilated stumps in despair from their deepening graves; —and beside these are others which have kept their feet with astounding obstinacy, although the barbarian tides have been charging them for twenty years, and gradually torn away the soil above and beneath their roots. The sand around—soft beneath and thinly crusted upon the surface—is everywhere pierced with holes made by a beautifully mottled and semi-diaphanous crab, with hairy legs, big staring eyes, and milk-white claws;—while in the green sedges beyond there is a perpetual rustling, as of some strong wind beating among the reeds: a marvelous creeping of "fiddlers," which the inexperienced visitor might at first mistake for so many peculiar beetles, as they run about sideways, each with his huge single claw folded upon his body like a wing-case. Year by year that rustling strip of green land grows narrower; the sand spreads and sinks, shuddering and wrinkling like a living brown skin; and the last standing corpses of the oaks, ever clinging with naked, dead feet to the sliding beach, lean more and more out of the perpendicular. As the sands subside, the stumps appear to creep; their intertwisted masses of snakish roots seem to crawl, to writhe—like the reaching arms of cephalopods. . . . Grande Terre is going: the sea mines her fort, and will before many years carry the ramparts by storm. Grande Isle is going—slowly but surely: the Gulf has eaten three miles into her meadowed land. Last Island has gone!

301

Lafcadio Hearn was right—Grande Isle *is* going! But it has not been ruined by the sea. It is the kingdom of oil that guts her beaches. We long now for those islanders that could give voice to islands that have been changed or threatened by man's inability to accept what he has. Listen to Walt Whitman tell you about his island—and then painfully, unmercifully slowly crawl down the Long Island Expressway.

Worth fully and particularly investigating indeed [is] this Paumanok, (to give the spot its aboriginal name,) stretching east through Kings, Queens and Suffolk counties, 120 miles altogether—on the north Long Island sound, a beautiful, varied and picturesque series of inlets, "necks" and sea-like expansions, for a hundred miles to Orient point. On the ocean side the great south bay dotted with countless hummocks, mostly small, some quite large, occasionally long bars of sand out two hundred rods to a mile-and-a-half from the shore. While now and then, as at Rockaway and far east along the Hamptons, the beach makes right on the island, the sea dashing up without intervention. Several light-houses on the shores east; a long history of wrecks, tragedies, some even of late years. As a youngster, I was in the atmosphere and traditions of many of these wrecks—of one or two almost an observer. Off Hempstead beach for example, was the loss of the ship "Mexico" in 1840. (alluded to in "the Sleepers" in *Leaves of Grass*) And at Hampton, some years later, the destruction of the brig "Elizabeth," a fearful affair, in one of the worst winter gales, where Margaret Fuller went down, with her husband and child.

The eastern end of Long Island, the Peconic bay region, I knew quite well too—sail'd more than once around Shelter island, and down to Montauk—spent many an hour on Turtle hill by the old light-house, on the extreme point, looking out over the ceaseless roll of the Atlantic. I used to like to go down there and fraternize with the blue-fishers, or the annual squads of sea-bass takers. Sometimes, along Montauk peninsula, (it is some 15 miles long, and good grazing,) met the strange, unkempt, half-barbarous herdsmen, at that time living entirely aloof from society or civilization, in charge, on those rich pasturages, of vast droves of horses, kine or sheep, own'd by farmers of the eastern towns. Sometimes, too, the few remaining Indians, or half-breeds, at that period left on Montauk peninsula, but now I believe altogether extinct.

More in the middle of the island were the spreading Hempstead plains, then (1830–'40) quite prairie-like, open, uninhabited, rather sterile, cover'd with kill-calf and huckleberry bushes, yet plenty of fair pasture for the cattle,

302

The bayou swamp-forest of Louisiana. (Courtesy Standard Oil Co. of N. J.)

mostly milch-cows, who fed there by hundreds, even thousands, and at evening, (the plains too were own'd by the towns, and this was the use of them in common,) might be seen taking their way home, branching off regularly in the right places. I have often been out on the edges of these plains toward sundown, and can yet recall in fancy the interminable cow processions, and hear the music of the tin or copper bells clanking far or near, and breathe the cool of the sweet and slightly aromatic evening air, and note the sunset.

Through the same region of the island, but further east, extended wide central tracts of pine and scrub-oak, (charcoal was largely made here,) monotonous and sterile. But many a good day or half-day did I have, wandering through those solitary cross-roads, inhaling the peculiar and wild aroma. Here, and all along the island and its shores, I spent intervals many years, all seasons, sometimes riding, sometimes boating, but generally afoot, (I was always then a good walker,) absorbing fields, shores, marine incidents, characters, the bay-men, farmers, pilots—always had a plentiful acquaintance with the latter, and with fishermen—went every summer on sailing trips—always liked the bare sea-beach, south side, and have some of my happiest hours on it to this day.

As I write, the whole experience comes back to me after the lapse of forty and more years—the soothing rustle of the waves, and the saline smell —boyhood times, the clam-digging, barefoot, and with trowsers roll'd up— hauling down the creek—the perfume of the sedge-meadows—the hay-boat, and the chowder and fishing excursions;—or of later years, little voyages down and out New York bay, in the pilot boats. Those same later years, also, while living in Brooklyn, (1836–'50) I went regularly every week in the mild seasons down to Coney Island, at that time a long, bare unfrequented shore, which I had all to myself, and where I loved, after bathing to race up and down the hard sand, and declaim Homer or Shakespeare to the surf and sea-gulls by the hour.

303

Yes, Long Island has changed; Grande Isle is awash with oil. What has happened, what *is* happening, to our island world?

17

What Has Happened to Our Islands?

NOT TO KNOW an island, says the Scottish poet and nationalist, Hugh MacDiarmuid, "is like having a blunt sensation in the tips of your fingers. Horrid! But to know a whole lot of islands is like having a portfolio of pictures and an adjustable frame, which enables you to hang up any picture for a day, a week, or a month. . . . Above all they are useful nowadays because an island is an almost startling entire thing, in these days of the subdivision, of the atomization, of life."

MacDiarmuid admits he is a passionate lover of islands. So are we. You may not have found in this book the island you most love. Perhaps that is a good thing. It will remain for a while longer *your* island. An almost hysteria is plaguing our islands, and island myths are daily created in the advertising media: You are urged to give your wife an island for Christmas; you are inveigled into hitching your star to a sea gull; you are told sharply not to keep your shoes on, but to take them off and race down the tropical sands by the waves' edge. Do it. But go carefully. There are sea urchins in the clearest water—serpents even in tropical Edens.

MacDiarmuid once sang about one of his islands:

"Do not be deluded! There is nothing there but just a lot of
water and rocks.
Just a lot of water and rocks—and peace and beauty and
the glories of an ancient people."

So do not be deluded! Many islands have known no glories; many are being despoiled of all beauty—and peace.

Shortly before World War II Katharine Fullerton Gerould said:

When the plight of the planet becomes desperate, people usually begin to babble about islands. I have recently been almost deafened with the word in the literature and conversation of escape; islands have always been more popular than caves or vales or mountain summits. They seem to spell the only secure isolation. Even when authors have merely wished to display some human experiment unhampered by society—Defoe, or Bernardin de Saint-Pierre, or Herman Melville, or H. G. Wells—they have chosen islands for their scene; and conversely, if authority wishes to be quite free of some human being it does not dare to sell, it finds an island for his exile; the Salt Islands—or L'Ile du Diable—or St. Helena.

The plight of our planet has always been somewhat desperate, but we think it important now to become somewhat desperate about our islands—many slippery with oil, many overdeveloped and distinctly misunderstood.

Suppose, as MacDiarmuid suggests, we frame some islands and see what has happened.

The islands of the world are having imposed upon them by builders and constructors a kind of terrible similarity. When the tourist associations of the world discovered the appeal of islands they discovered a deep need within us and, as with most needs, once it is discovered it is shamefully exploited. If our sailing ships once brought disease to the South Seas, our planes now bring a new form of parasitic growth—the developer. You can now buy an island in the South Pacific through such people as Robert Harold Hunter, who says that "islands are my thing." He has sold some 300 islands and admits to having done more than $1,000,000 worth of business in Fiji alone. "I can fly over an island," says Hunter, "and feel its potentialities; looking down, I analyze where the coral line is, whether the coral is alive or dead, whether you can swim there at high or low tide, where you can anchor your boat, whether it is possible to comb the jungle out of the palm trees."

Robert Hunter is a good man, and indeed an island lover, but with the growing disenchantment in the Pacific, are marinas the answer?

The changes today in the Pacific islands are infinite—from the new nationalism in the Philippines to the misery in Micronesia there is a new passion upon the coral atolls: Those idyllic pictures, for example, of breadfruit, the ever-present coconut, the luaus, all are overshadowed by the fact that many islanders are just plain hungry. More than two decades ago we evacuated the islanders of Eniwetok and Bikini, both in the

305

Alligators in a Southern bayou. (Nineteenth-century print from the authors' collection)

306

Skin diver glides above swarms of gaily colored reef fish in a forest of elkhorn coral near Buck Island, Christiansted, in the Virgin Islands. (National Park Service photograph by M. W. Williams)

Marshall Islands, a part of the Trust Territory of the Pacific Islands, a United Nations trusteeship under United States administration. Eniwetok had some forty islets. The islanders used the lagoon—some 388 square miles—not as a marina but as a marine garden which produced their food. They were evicted to the atoll Ujelang, which has only thirty-two islets, a land area far too small to feed them, and a lagoon only one sixteenth of the size they left behind. We sent the natives of Bikini to an island, Kili, which has no lagoon at all, something which the islanders simply cannot understand, having a whole culture built around lagoon life. In addition Kili is surrounded by rough seas, and ships touch upon it only every four months while those peoples evacuated from Eniwetok are completely off the normal trade route of the Marshall Islands. It is true that the coconuts and the fish are still available, but as throughout all the islands, the diet of the peoples has changed; they long for some imported canned goods; they have grown used to imported rice. They wait impatiently for the day when there will no longer be any radioactivity on Eniwetok and Bikini, so they may return.

All of Micronesia is plagued with problems; it is an area that we have ignored too long. Whether it will eventually become associated with the United States in a more intimate fashion, or whether it will become independent, is something to be studied and a plebiscite has been urged for 1972. It is a far-flung territory, situated in ocean seas as large as the United States, but in terms of just plain land it is only about half the area of Rhode Island. The islands have known many peoples—the Spaniards, the Germans, the Japanese, and now the Americans. Some call the islands the most beautiful scrap heaps in the world. There are leftover quonset huts from the war in the Pacific, tin towns, poor roads or none. Our government has never authorized funds for such things as the permanent construction of roads, or for a territorial vocational school, or for new hospitals and new machinery. We *have* sent Peace Corps volunteers, many who speak one of the nine major languages, many who are building their own schools out of coral blocks, or working as X-ray technicians, civil engineers, nurses, business specialists, and radio operators.

Or let us frame Australia's famous Great Barrier Reef, which is being devoured—and distinctly by the ignorance of man.

The beautiful giant triton, one of the most glorious of all shells, has long attracted the eye of the collector. In years gone by it had a useful purpose as a trumpet that South Sea islanders used for signaling from island to island. A foot in length, handsome, it is a gastropod mollusk that is a predator, but not nearly the predator that the starfish, the

307

308 A beach that is part of the Great Barrier Reef. (Courtesy Australian News and Information Bureau)

Crown of Thorns, is. The starfish with a Gargantuan hunger is literally eating up the corals of the reef. Studies have been taken to show where the damage is most acute and finally the Australians decided that it was only in those areas where there was heavy tourism—which encouraged triton shell collecting—that the infestation was common. A 300-mile stretch of reef on the northeast coast of Australia has been severely hit. On the accessible side, ninety percent of the coral has been killed. When the coral dies, the waves take over and the reef disappears. A plague of starfish has been attacking all the Pacific islands: Guam, Rota, Johnson Island, Fiji, Truk, the Palaus as well as the Hawaiian Islands. It is more than likely that radioactive fallout and pesticides have also contributed to the destruction of reefs. But it is also apparent, at least to the Australian students of the reef, that the triton, which feeds on the starfish, is also disappearing and that the balance of nature has once again been destroyed. Coral, which in some colonies takes more than a hundred years to grow, can be devoured within a very short period of time.

The great natural resources of the Chincha Islands, the bird islands of Peru, are disappearing too. These guano islands are world-famous.

Gathering coral on Australia's Great Barrier Reef, the great bastion of coral which extends for 1200 miles down the coast of Queensland from the tip of Cape York Peninsula to the region of Rockhampton. (Courtesy Australian News and Information Bureau)

Guano is one of the world's greatest fertilizers, the nitrogen-rich excrement of the cormorants—handsome, long-necked seabirds, larger than the gull. Guano harvests have always been important to Peru's economy; indeed the ancient Incas discovered these offshore guano islands, and many ritual remains have been buried under the bird droppings. The nineteenth century was a time of almost wholesale exploitation of the guano fields and more than $3,000,000,000 has emerged as profit from the guano. Indeed, it was not the gold and silver of the Incas that were Peru's hidden wealth but the fertilizer product of these islands.

The guano, however, is disappearing; last year's harvest was down to 35,000 tons, a drop from 170,000 tons the year before. The answer? The birds are starving; the balance of nature has been upset. It is a simple matter, stemming from the production of fish meal, now one of Peru's largest exports; but fish meal is made from anchovies, and anchovies are the food of the cormorant. The cormorant may disappear, but equally important the anchovy will disappear too. As Felipe Banavides, a Lima businessman and conservationist, says, it is overfishing of the worst sort because the guano that is deposited in the sea provides plankton and it is

Maintaining offshore drilling rigs requires a fleet of boats to ferry supplies and men. (Courtesy American Petroleum Institute)

One of a series of twelve crystal plates engraved to depict collectors' shells gathered from the seven seas. The shell depicted here is the trumpet triton. (Courtesy Steuben Glass)

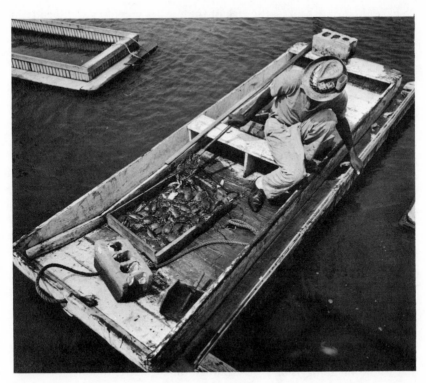

311

Maryland blue crabs. Crab production is a paramount industry on Smith Island. Here a man dips crabs into his skiff from a crab box where the crabs were raised under controlled conditions. (Photograph by M. E. Warren)

the plankton that the anchovies eat.

What about framing islands closer to home?

Just about fifty miles from Times Square is an island chain, a little archipelago called the Norwalk Islands in Connecticut. The story goes that a few years ago some Sunday sailors in a large power cruiser saw these islands, which have been important to us since childhood, and stated, "My God, do we have something like this close to Manhattan?" If one does, one obviously has to change it. Plans were set up, plans that are becoming uniform around the world: a giant marina, luxury high-rise real-estate developments. Fortunately many banks are still suspicious of islands, and the financial requests were not met; the plan was abandoned. The islands were not, perhaps, big enough or important enough for a bank, or maybe the banker was simply a good New Englander who would not dip into our coastal capital—our islands. The Norwalk Islands do have their drawbacks—the lack of fresh water was often a difficulty in the past.

Cockenoe Island was saved, thanks to spirited citizens, from having a nuclear-powered public-utilities plant placed upon it. Of course we need power, but we need islands too. These islands for all their size have had a varied history, having known many island lovers—the Indians and at one time a group of Mormons. They have known older oystermen who are now being threatened with the loss of the oyster beds; old captains; Connecticut children of many generations, and occasionally such celebrities as Lillian Hellman and Billy Rose, who built houses on the chain. The community is now eager to save its islands and it is amazing with what concerted efforts such community pressure can be exerted upon island development. When Fire Island off Long Island was about to be paved by Robert Moses the voice of the people was distinctly heard in the land and eventually the island became a National Seashore preserve. Many of our islands are now protected at least from wholesale development by being made part of the National Seashore or National Park systems.

But those people who have been islanders for years are by their very nature not knowledgeable of the worlds they possess. Many islanders have been extremely poor conservationists—a pattern that went back to the development of our country when such great tracts of woodland as that found on Long Island were burned off regularly, following the old practice of the Indians. And although good oystermen and crabbers and lobstermen, they have made little effort to understand the resources surrounding their enviable pieces of land. Dredging of oysters has meant a steady decline in many areas of our country, for example, particularly around Smith Island and Tangier in the Chesapeake, once almost completely isolated. Those islanders spoke a language of Chaucer's time and the story goes that one Smith Islander who went to New York returned to his waiting relatives and was asked his impressions of that other great island, Manhattan. He said, "Well, it'll never amount to much, it's too far away."

Even the old Fire Islanders, the few that there were, and those along the barrier islands in areas from Georgia to Long Island, have mistreated the land. Of course nature would sometimes start that destruction. Just a few years ago violent storms with winds at sixty miles per hour whipped high spring tides across a thousand miles of ocean. Our barrier islands on the East Coast were flooded. New Jersey had 2400 houses destroyed, 8300 damaged and $80,000,000 worth of direct damage in just three days of storm. Man blamed nature. The old islanders said about the storms

312

One of the Guano Islands off the coast of Peru. (Courtesy Grace Line)

313

Arctic angler tries her luck through the winter ice near King Island Village at Nome, Alaska. (Courtesy Alaska Travel Division)

that God brought the sand and God takes it away, but the truth is that man's ignorance of the sand is the true culprit.

Dunes, of course, must be stabilized. In some countries perpetually attacked by water, the Netherlands for example, man has had to learn the nature of an island, but largely we have remained abominably ignorant. Dunes have been breached; home have been built where none should have been allowed; vegetation has been allowed to die or its growth inhibited by removal of ground water; parking areas have interfered with the recharge of aquifers; we have made our own islands vulnerable. The storm, of course, could not have been averted, but some of our island beaches are remarkably self-stabilizing; dunes can remain protectors, sentinels. Real-estate developments should never be permitted; the inevitable parking areas, even in undeveloped sections, should be made of flexible materials. Our coasts have always been overdeveloped, for we are a country which leads itself to beaches, be they island or mainland. Sometimes the enemy of our natural environment is simply the car.

Islands, even the smallest of them, take care of themselves if left alone. They are extremely happy with their natural vegetation—the coral cays of the Caribbean are an example. The pioneer strand of the plant community was sown with seeds brought in by the sea or by migrating birds. The result was a plant community of low scrub going to a climax stage of broadleaf woodland. Man came, cut down the woods and frequently planted the wrong thing—coconuts, for example, which we think indigenous to these islands. They were, in fact, not. Recently during a hurricane in which winds exceeded two hundred miles per hour the sea rose fifteen feet above its normal range. These coral cays are just ten feet above sea level and they were completely inundated. A study was made to see what damage was done to the vegetation cover. On those islands which had been cleared and replaced with coconut plantings the stabilizing undergrowth had been removed. The waves tore up the trees, denuded the islands. Those islands that had been untouched adapted to the terror of the sea, rootholds were kept, erosion was limited and indeed the cays grew rather than eroded, sometimes rising as much as five to seven feet. Throughout this area at least twenty good-sized islands have actually disappeared because of the introduction of that excellent export crop—the coconut—in the last half of the nineteenth century. Now we are planting not just the coconut, but the tourist. But happily some of these cay islands are reverting to their littoral thicket, and a new equilibrium might be on the way.

Even New York may make some effort to preserve its islands—Staten Island, for example, beloved by Thoreau and beautifully recorded by his friend William T. Davis. "There is no need," he said, "of a far-away fairyland, for the earth is a mystery before us. The cowpaths lead to mysterious fields."

The cowpaths of Staten Island have disappeared, but Davis can still remind us:

A man who concerns himself principally with the artificial, and who thinks that the world is for stirring business alone, misses entirely the divine

314

Snorkel divers get ready to explore the translucent waters of Australia's Great Barrier Reef, an underwater miracle of color fringing the northeast coast for 1200 miles. (Courtesy Australian National Travel Association)

315

Divi-divi trees, shaped by the steady trade winds, are an unforgettable feature of Aruba, off Venezuela. (Courtesy Aruba Tourist Bureau)

halo that rests about much in nature. To him all things are certain. He can have a particular tree cut down or an ox killed at command. You see him hurrying across the street with rapid strides, for hasn't the Valley railroad declared a dividend! Such things must be, but they are not the safest springs of pleasure. We must not put by entirely the chippy singing in the apple tree, or the white clouds, for nature declares a dividend every hour.

Once a beautiful rural island, Staten Island has been changed by a bridge. The island, says Lorin McMillen, Director of the Staten Island Historical Society, began developing steadily after World War II, but the Verrazano Bridge has made the island boom in these past few years more than during the thirty years before. The statistics about its popula-

ISLANDS

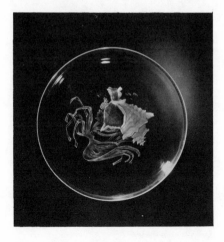

Banded murex: One of a series of
twelve crystal plates engraved to depict collectors' shells gathered from
the seven seas. (Courtesy Steuben
Glass)

316

tion growth in that period are extraordinary. Since the opening of the
bridge on November 21, 1964, the population has increased from 60,000
to 310,000. There are, of course, still some old communities and many
families have lived for well over five generations on the island. Such
spots as Travis, Rossville, and Charleston remain almost the same as
they were, but still they complain. "This will be worse than Long Island"
is the usual comment. But Staten Island still has its parks, and the islanders are determined to preserve them. The Staten Island Citizens' Planning Committee and the Staten Island Greenbelt Natural Areas League
(SIGNAL) are making a real effort to preserve the natural island landscape. There are proposals now for:

Creation of a new commission-level agency to resolve land-use conflicts
arising between existing city agencies and to protect the city's priceless park-
lands from municipal encroachments.

Appointment of an assistant administrator or commissioner of conservation within the Parks, Recreation and Cultural Affairs Administration.

Compilation of an inventory of ecologically significant natural landscapes within the city, the establishment of a program to preserve these areas
permanently as natural landmarks.

A sustained effort by the Boards of Education and Higher Education to
integrate environmental lessons, field trips and programs into their curricula.

Not only the parks themselves, but other exciting areas of New
York, must be saved—for example, the beautiful natural marsh area in
the Mill Basin section of Brooklyn near Jamaica Bay. For those who are
always inspired by a marine landscape, no matter how defiled by abandoned automobiles and garbage dumps, the Mill Basin area still has an

excitement about it. A sanctuary for ducks and a spawning ground for fish, it is nonetheless polluted. But things could change and there are people who want them changed. On one side are the conservationists who maintain, "Everyone is always worried about saving faraway forests or isolated islands, but what about this place that people can appreciate right here?" And indeed, resting as it does within New York City, it is something to save. But will it be? On the other side there are the developers, one of whom says, "They say it's a place for bluefish to spawn; phooey, it's a place for rats!" What he plans to offer instead is one of the glorious shopping centers that disfigure our landscapes, but which are, as he puts it, "a star in the firmament."

There is a new movement in our land to avoid the ecological suicide that our policy of self-destruction has been encouraging. The voices are there, if we will only listen to them; from New York to California, island ecologies are being reevaluated. The regents of the University of California are making an earnest effort to undertake such ecological studies. Dr. Mildred E. Mathias, a botanist at the University of California, maintains, "It is from this field of study that we hope to find the key to man's survival and his relations to an environment which is becoming increasingly altered under the pressures of urbanization and population increase."

317

Ano Nuevo Island (off the California coast), for example, is now used as an area to study animal behavior. On its eight acres it supports five different kinds of seals and sea lions, half a mile from shore. We must now frequently move away from the mainland to study animal life. Islands have not been so badly altered from the natural environment that they cannot be the laboratories that will teach tomorrow a way to look at the world that we ignored yesterday.

Islands in all coastal areas have a range of natural and seminatural communities, but in addition the changes and evolution of such areas are extremely rapid. Therefore in terms of studying natural phenomena they offer unrivaled opportunities. And some can be saved; many of our marsh islands, for example, can be reclaimed.

Marsh islands, just like beaches, are easily eroded. They need plant life as do the vulnerable dunes of the beaches. Many of us have seen marsh islands grow and die before our eyes, have seen sediment settle over rocks or sand flats. Some was washed away in the tide, but other accumulations become more stubborn. Seeds began to sprout and the deposition of more silt and clay continues until a true tidal island develops that is no longer completely at the mercy of the tides that inundate it.

The conservation movement in the United States has grown somewhat as a tidal island does, bits of sand growing on the stubborn rock of man's conscience. It will no longer be inundated by the overwhelming tides of technological progress. In conservation we are conserving and preserving ourselves, not, as Lewis Mumford remarks, out of concern for a threatened species—a whooping crane, a tidal island, a mussel flat, a tree—but because there is some wildness in us that must be protected. "We must know what we are doing," he says. "We are rallying to preserve ourselves, we are trying to keep in existence the organic variety, the whole span of natural resources upon which our own development will be based. If we surrender this variety too easily in one place we shall lose it everywhere; and we shall find ourselves enclosed in a technological prison without even the hope that sustains a prisoner in jail—that someday we may get out. Should organic variety disappear, there will be no 'out.'"

And for that reason, for many of us there must always be islands.

Appendix

Sailors since the time of the Polynesians have always approached islands with discretion. These observations are from: *Wind Waves at Sea Breakers and Surf,* by Henry B. Bigelow, Museum of Comparative Zoology, Cambridge, Massachusetts; and W. T. Edmondson, Woods Hole Oceanographic Institution, Woods Hole, Massachusetts. Courtesy of Hydrographic Office, H. O. Pub. 602, Washington, D.C.

SURF AROUND ISLANDS

The factors that determine the differences in heights of the breakers from place to place along coasts in general, act in the same way around the shores of islands, whether large or small. Thus, the waves are likely to be focused, as it were, on the exposed side of an island that is rounded in outline if it rises from water shoal enough to alter the direction of advance of the waves to any considerable degree. Consequently, just as at the tip of a headland, a worse surf may be expected there with a given wind than would develop on a straight coast. But while the waves are often refracted right around a small island of this shape, the heights of the breakers they produce will decrease following around the shore, as their inshore ends are delayed more and more by the effect of the bottom. Theoretically, the inshore ends of waves, that were initially 20 to 100 times as long as high, should break at angles of about 16° to 35° with the part of the shore that was at right angles with their crests offshore, assuming that they did so where the depth was equal to 1.3 times their own heights at breaking, while their heights would be reduced by a little more than one-half there accordingly, as compared with the most exposed part of the island. Observations suggest that the reduction in height might actually be of about this general order of magnitude. Theoretically, too, the inshore ends of the waves should lose still more in height by the time they reach the more sheltered side of the island. However, landing is not apt to be as much easier there as this might suggest, especially if the island is small, for the following reasons:

a. Since the inshore ends of the waves are still at a considerable angle with the coastline when they break, and since the coasts of round islands are usually rocky, bouldery, or strewn with coral heads, landing is much more difficult for practical reasons, than it would be if waves of equal height were breaking parallel with the shore, and if the latter were sandy or pebbly.

b. Although the reduction, by refraction, in the heights of the waves around a small circular island is greatest on the most protected side, the surf may be made very confused there because of the interference that often develops between the two trains of waves that meet, as they are refracted around from the two sides. We ourselves have vivid memories of attempts to land on rocky islets that were unsuccessful for this very reason.

In short, the chance is not very good of landing anywhere around the shore lines of a small rocky island that is circular in form, if the sea is too heavy to allow this on one or other of its two lateral quadrants, and if the submarine contour is such that the waves are refracted right around it. But there may be a shadow zone of quiet water in the lee of an islet, if its shores rise abruptly from water so deep that the waves then running are refracted but little as they approach it, just as there may be in the lee of a promontory of similar character. But anyone who takes advantage of this to land will be well advised to keep a sharp eye on the state of the sea, and be prepared to put off again at once if the latter rises, for a troublesome surf may develop with astonishing suddenness, as we have often seen.

The more irregular the coast of an island is, and the more abruptly it alters from place to place, the more likely it is that one can find a place in some cove, or in the lee of some headland that will be sheltered from the surf, under conditions of sea that would prevent landing on the more exposed parts of the shore. One or the other of two shallow bights, for example, marked A and B on figure 1, that flank a short promontory on the northern side of No Man's Land, off Martha's Vineyard, is usually sheltered enough from southerly swells for landing, except in really heavy weather, although the island is only about 1½ miles, east and west, by about 1 mile, north and south, with a very even shoreline, and without any offlying shoals or reefs to break the seas. Similarly, when coaling from one steamer to another was impossible in Cook Bay on the southeast side of Easter Island off the coast of Chile in the third week of December, 1904, because of a heavy swell from the southwest, we found La Pérouse Bay on the northeast side so protected by Cape Roggewein on the one hand and by North Cape on the

320

other side, that the two steamers could lie side by side in the open roadstead.

A cove on the leeward side of even a small island may, indeed, be perfectly sheltered if the shore line is broken up by a succession of headlands, especially if there are offlying ledges or islets to interfere with the wave pattern. The harbor on the east side of St. Pierre Island off the south coast of Newfoundland (fig. 2) is an excellent example, being well protected in this way from swells from the southward, southwestward, and southeastward, although it is wide open to the northeast, and although the distance from its entrance to the most easterly promontory of the island is only about 1½ miles.

With respect to islands that are much longer than they are broad, especially if they are linear in form, the sheltered side of an island of this shape, lying athwart the general run of the waves, may offer excellent anchorage and easy landing, if it is more than a few hundred yards long. A long, narrow crescentic island with the concavity on the sheltered side, as is true of many sandy islands in regions where the strongest winds are prevailingly from one direction, acts still more efficiently as a natural breakwater. However, if the waves are striking it endwise, no shelter can be expected on either side of it. Sable Island, off Nova Scotia, which is somewhat crescent-shaped, with its main axis easterly and westerly and with its concavity on the northern side, provides shelter in this way from southerly swells or seas; but there is no shelter anywhere around its shores if the wind is from any other quarter, in spite of its considerable length (21.5 miles), and even though an extensive bar makes out from each end of it.

321

The larger an island is, the lower the surf is on its most sheltered side, as a rule, partly because the sidewise expansion of the inner ends of the wave crests is greater around it than in the case of a smaller island, but also because the interference by irregularities of the coast, operating through a longer distance, drains the inshore ends of the waves of their energy more effectively.

We might also remark, in passing, that the presence of an island, of whatever size or shape, may interfere with the regular wave pattern for a long distance to leeward, in regions where long swells prevail. The Polynesian navigators of old were acquainted with this phenomenon, and made use of it, not only to direct their canoes from island to island over long distances, but also in their search for new islands.

The preceding account of surf around islands in general, and around promontories, applies equally to coral atolls, for these are not all circular in shape, as is the common belief, nor even approximately so, but exhibit a wide variety of outlines. Thus, the swell heaves right around Nukuoro Atoll, in the Carolines, during the winter season when the northeast trade winds are at their height, there being no shelter anywhere, except within its entrance, for it is nearly round and only between 3 and 4 miles in diameter (fig. 3). On the other hand, the southerly face of Arno Atoll is so well protected by the long promontory known as Northeast Point that it affords a safe anchorage and easy landing when the swell is coming from the northeast, though not with swells from any other direction.

Waves that run in through the passages by which the lagoons of coral atolls are connected with the open sea expend their energy in the basins inside, at the expense of their heights, just as happens in harbors of similar shape. And it is well known to everyone who has had experience in coral seas that there is no danger from surf once one is inside the lagoon, no matter how heavy the breakers may be on the exposed faces of the reefs outside, provided only that the contour of the lagoon falls away sharply inside the entrance. If only the one side falls away, this will be the more quiet side, the other the least so. The measurements of waves that have been made in Duluth Harbor suggest, for example, that sidewise expansion would not only reduce waves that were 15 feet high outside to only about 2 feet high at a distance of a quarter of a mile inward from the entrance of an atoll as nearly circular as the one charted in figure 3, but would render them imperceptible by the time they had crossed the lagoon, if the latter were so much as one-half mile across. And the waves running in through the passage would be drained of so much of their energy as they broke along its sides that their heights would already be reduced considerably before they reached the lagoon at all.

322

A Little Treasury of Islands

There are over 500,000 indexed islands; this small sample can but whet the appetite.

Adak Island: Located in SW Alaska, this island is part of the Andreanof group, Aleutian Islands. It is 30 m. long and its width varies from 3 to 20 m.

Adelaide Island: 59½ m. in length and 17½ m. wide, this island is just off the west coast of Palmer Peninsula in Antarctica.

Admiralty Island: A large island in the Alexander Archipelago of SE Alaska, it covers 1664 sq. m.

Aegean Islands: These islands are all parts of an old submerged block of land, and their chief products are coral and sponges. Located off Asia Minor, they comprise 1506 sq. m.

Aegina: In ancient times this island was of great commercial importance until its defeat by Athens in 459 B.C. The first Greek coins were minted here. It is located in the Saronic Gulf, E central Greece, and is 32 sq. m. in area.

Afognak Island: In southern Alaska, this island is located between the Gulf of Alaska and the Shelikof Strait. It is 9–23 m. wide and 43 m. in length.

Aland Islands: Finnish archipelago in the Baltic Sea. In all, this group totals 572 sq. m. and has 6500 islands and rocks. Eighty of the islands are inhabited and Aland is the largest with 285 sq. m. area.

Alcatraz: Former U.S. Federal prison island in W. California's San Francisco Bay.

ISLANDS

Aldabra Island: In the Indian Ocean, it is the largest atoll of the Seychelles. It is 21 m. long and 8 m. wide.

Alexander I Island: Discovered by Thaddeus von Bellingshausen, it was named for the Russian tsar. Approximately 45–85 m. wide and 205 m. long, it is located in the Bellingshausen Sea, Antarctica.

Alexandra Land: In the Arctic Ocean, this is the westernmost island of Franz Josef Land. The most northerly land in the eastern hemisphere, it is 70 m. long and 10–30 m. wide.

Alor Islands: This group of the Lesser Sundas in Indonesia actually consists of two islands comprising 1126 sq. m. They are Pantar and Alor (Ombai). The latter has area of 810 sq. m.

Als: This island in Denmark's Little Belt had originally belonged to Germany until the plebiscite in 1920. It is 121 sq. m. in area.

Alsten Island: This island's chief attraction is its Seven Sisters peaks, which are celebrated in legend. This rugged island, situated in N Central Norway, is 59 sq. m. in area.

Amager: This island includes part of Copenhagen in its northern part. It is 25 sq. m. in area and its location is Oresund, Denmark.

Amakusa Islands: A total of 70 islands and islets, this group is found in the East China Sea, Japan. Small islands, they total only 341 sq. m. in area.

Ambergris Cay: Part of north British Honduras in the Caribbean Sea, this island is 5 m. wide and 25 m. in length.

324

Amboina (Ambon): Site of the "Massacre of Amboina" in which the Dutch nearly annihilated English merchant settlers on the island. Later this island in the Moluccas, Indonesia, opposed Indonesian independence, leading to a revolt in 1950. It is 314 sq. m. in area.

Amchitka Island: One of the Rat Islands in SW Alaska, this 40 m. long and 2–5 m. wide island was the site of a U.S. air base during World War II.

Amin Divi Islands: The northern group of the Laccadive Islands in the Arabian Sea, India, their chief products are copra and fish.

Amirantes (Amirante Isles): An archipelago in the Indian Ocean and an outlying dependency of the Seychelles, it is 100 m. long N–S.

Anambas Islands: One of the most important groups of islands in Indonesia's South China Sea, it is comprised of three islands, the largest of which is Jemaja (15 m. long and 10 m. wide).

Andoy (Anda, Anna): Within the Arctic Circle, this 150 sq. m. island is the northernmost of the Vesteralen group in N Norway.

Andreanof Islands: In Southwest Alaska, these islands comprise one of the main groups of Aleutian Islands.

Andros: The northernmost island of the Greek-owned Cyclades in the Aegean Sea, it is 145 sq. m. in area.

Andros Island: The largest of the West Bahama Islands (1600 sq. m.), it has the only river in the Bahamas.

Angel de la Guarda Island: One of the larger islands in the California Gulf, NW Mexico (330 sq. m.), it is barren and uninhabited. It is used seasonally as a fishing base.

Kapok tree growing in Virgin Islands National Park. (Courtesy National Park Service)

Angel Island: 1½ m. in length, it is the largest island in the San Francisco Bay. It has variously fulfilled the functions of a quarantine station, an immigration station, and an enemy internment camp.

Anglesey (Anglesea): A 275 sq. m. island in the Irish Sea, NW Wales; two bridges connect it to the mainland.

Anjouan Island (Johanna): A major island of the Comoro Islands, it was formerly a notorious slave market. It is located in the Mozambique Channel and is 138 sq. m. in area.

Annette Island: This 20 m. long and 7–10 m. wide island is a Tsimshian Indian reserve in SE Alaska's Gravina Islands.

Anticosti Island: A large island and the source of much lumber, this island is situated in the Gulf of St. Lawrence, East Quebec, and is 140 m. long and 30 m. wide.

Antigua: One of the Leeward Islands, this 180 sq. m. island's character is peculiarly "Scottish." Its primary products are sugarcane and sea-island cotton.

Anvers Island (Antwerp): This 34 m. long and 25 m. wide island arises to a height of 9412 feet. It is the largest of the Palmer Archipelago in the Antarctic.

Aoba (Leper Island): New Hebrides island of volcanic origin, it is 95 sq. m. in area.

Aran Island: A lighthouse island off the W coast of Ireland's County Donegal on the Atlantic Ocean. It is only 4 m. long and 3 m. wide.

Arran: This island in the Firth of Clyde, Scotland, is 166 sq. m. in area and has an abundance of deer.

Assateague Island: This 35 m. long narrow barrier island off the Atlantic shore of Maryland and Virginia is now under the jurisdiction of the Department of the Interior.

ISLANDS

Assumption Island: Of coral origin, this island is one of the Aldabra group and an outlying dependency of the Seychelles.

Atka Island: The site of the Korovin Volcano, which rises to a height of 4852 feet, it is one of SW Alaska's Andreanof Islands and is 65 m. long and 2–20 m. wide.

Attu Island: A rugged, barren island occupied by the Japanese in World War II, it is the largest of the Near Islands and the westernmost of the Aleutians. It is 30 m. long and 8–15 m. wide.

Australia: The largest island or the smallest continent, it has an area of 2,948,366 sq. m. and is located SE of Asia.

Austvagoy: Largest of Norway's Lofoten Islands in the Norwegian Sea, it is 203 sq. m. (33 m. long and 11 m. wide) in area.

Avery Island: One of south Louisiana's Five Islands, Avery houses a bird sanctuary and the Jungle Gardens, noted for rare plants.

Awaji-shima: The largest island in Japan's Inland Sea, it is 228 sq. m. in area and roughly triangular in shape.

Axel Heiberg Island: The largest of the Sverdrup Islands in the Arctic Ocean, it is 13,583 sq. m. (220 m. long and 20–100 m. wide) in area. Its coastline is indented by deep fjords.

Babar Islands: Consists of one large island (Babar) and five islets for a total of 314 sq. m. in the Banda Sea near Indonesia. It is a group of the South Moluccas.

Babuyan Islands: The major commodities are fish and rice. This island is part of the volcanic group in Luzon Strait, Philippines.

Bailey Islands: Formerly called the Coffin Islands, these islands are of volcanic origin and stretch for an 11 m. chain in the west Pacific. They are the southernmost group of the Bonin Islands.

Balleny Islands: 150 m. off Victoria Land, this glaciated volcanic group N of Antarctica was discovered in 1839 by the British sealer *Balleny.*

Banggai Archipelago: Consisting of Banggai, Peleng (largest island in the group), and c. 100 islets, it has a total area of 1222 sq. m. and is located in the Molucca Sea, Indonesia.

Bangka (Banka): A major producer of tin, this irregularly shaped island is 140 m. long and 70 m. wide. It is off the SE coast of Sumatra in the Java Sea and is part of Indonesia.

Banks Island #1: A 388 sq. m. island in Hecate Strait, West British Columbia.

Banks Island #2: Northwest Territories island in the westernmost part of the Arctic Archipelago (Arctic Ocean). It is 26,000 sq. m. in area.

Banyak Islands (Banjak): An Indonesian group of numerous small islands, the largest of which is Tuangku (Toeangkoe), 20 m. long and 6 m. wide.

Baranof Island: Formerly a headquarters for Russian fur-trading ventures, this 1607 sq. m. island is part of the Alexander Archipelago in SE Alaska.

Basse-Terre: The home of the dormant volcano Soufrière, this island is 364 sq. m. in area and forms the western half of Guadeloupe.

Bathurst Island: One of the Northwest Territories' Parry Island in the Arctic Ocean, it is 7272 sq. m. in area.

Batu Islands (Batoe): Comprised of many islets and three large islands for a total area of 463 sq. mi. This group is off the west coast of Sumatra, Indonesia, in the Indian Ocean.

Bay Islands: 144 sq. m. archipelago in the Gulf of Honduras, Caribbean Sea.

Beijerland (Beierland, Beyerland): Sometimes called Hoeksche Waard, this island is in the SW Netherlands, S of Rotterdam, and is 16 m. long and 9 m. wide.

Belcher Islands: 1096 sq. m. group of islands in the eastern part of Hudson Bay, Northwest Territories. These islands are the home of a small Eskimo population. The largest of these is Flaherty Island.

Bering Island: A 54 m. long and 22 m. wide island, it is the largest of the Komandorski Islands, off Kamchatka Peninsula in the SW Bering Sea. This is the island where the great Danish navigator Vitus Bering died in 1741.

Biak: Largest of the Schouten Islands, Netherlands New Guinea, it is 50 m. long and 25 m. wide.

Bickerton Island: An aboriginal reservation in the Gulf of Carpentaria, Australia, it is 12 m. long and 10 m. wide.

Bigge Island: A barren island composed almost entirely of quartzite, its location is the NE entrance to York Sound in the Timor Sea, Australia. It is 14 m. long and 7 m. wide.

Bijagos Islands (Bissagos): E Atlantic archipelago just off Portuguese Guinea, to which it belongs, it has fifteen large islands and several islets totaling 600 sq. m.

Billiton (Beliting, Belitoeng): A virtually square island of 1866 sq. m. area, it is the site of important tin mines. Its location is Indonesia, between the Java and South China Seas.

Bimini Islands: A string of cays in the Bahama Islands, this 8.5 sq. m. area is a popular tourist resort.

Boa Vista Island: This 239 sq. m. roughly circular island is the easternmost of the Cape Verde Islands. It was originally called São Christovao.

Bogoslof Island: Of fairly recent formation, this island first appeared in 1796 and its shape has been changed several times since by eruption. It is a volcanic islet in the Aleutian Islands, SW Alaska.

Bolshevik Island: 20 percent of this 450 sq. m. island is covered by glaciers. It is a southern island of the Severnaya Zemlya Archipelago in the Arctic Ocean.

Bonaventure Island: Small 2½ m. long island in the Gulf of St. Lawrence, E Quebec. A bird sanctuary, the island was granted to a Captain Duval by King George III.

Borden Islands: Two islands in the Arctic Ocean, Northwest Territories, they were formerly thought to be a single island. The two comprise 4000 sq. m.

Bougainville: Of volcanic origin and the largest of the Solomon Islands in

327

the SW Pacific, this mountainous island of 3880 sq. m. rises to 10,170 feet in Mt. Balbi and is governed by Australia under a U.N. mandate.

Boularderie Island: Off Cape Breton Island in NE Nova Scotia, this island is 2–6 m. wide and 25 m. long. It has valuable coal mines in the north.

Brabant Island: 29 m. long and 17½ m. wide, this is the second largest island of the Palmer Archipelago, Antarctica.

Brac Island (Brach): In the Adriatic Sea, this 152 sq. m. tourist center is the largest of Yugoslavia's Dalmatian Islands.

Brownsea Island: At the E end of Poole Harbour, SE Dorset, England, this 2 m. long and 1 m. wide island was the scene of the first camp of the Boy Scout movement.

Bruny: An island in the Tasman Sea, off the SE coast of Tasmania. North Bruny and South Bruny are connected by a narrow isthmus. Its area is 142 sq. m. (32 m. long and 10 m. wide).

Bubiyan Island: Belonging to Kuwait, this uninhabited island at the head of the Persian Gulf is 25 m. long and 15 m. wide.

Buton (Boeton): This island's chief product is asphalt. Its location is between the Flores and Molucca Seas, Indonesia, and it has an area of 1759 sq. m.

Bylot Island: This island, 4968 sq. m. in area, consists mainly of an ice-covered plateau. It is off Baffin Island in the Arctic Ocean, Northwest Territories.

328

Caicos Islands: These islands are a dependency of Jamaica and are the W group of the Turks and Caicos Islands.

Calamian Islands (Calamianes): This is a 600 sq. m. island group in the Philippines between Palawan and Mindoro.

Calvert Island: A 100 sq. m. island in Queen Charlotte Sound, SW British Columbia, it is 20 m. long and 2–10 m. wide.

Camagüey Archipelago: This archipelago consists of coral reefs off the N coast of E Cuba which extend approximately 150 m. NW–SE and form the southern flank of Old Bahama Channel.

Canarreos, Los: This archipelago off SW Cuba consists of a chain of numerous keys extending approximately 60 m. along the N and E shores of the Isle of Pines.

Capri: A rocky island in W Italy's Bay of Naples, it is famous for its grottoes and scenery and has an area of 5½ sq. m.

Catba Island (Cacba): A 15 m. long and 13 m. wide island in the Gulf of Tonkin off the N Vietnam mainland.

Cat Cays: A 60 m. long string of islets adjoining South Bimini, NW Bahama Islands.

Cat Island: This long and narrow island in the central Bahamas is said to be the most fertile of the archipelago. It is 50 m. long and 3 m. wide.

Cebu: This island in the Visayan Islands, Philippines, is volcanic but largely overlaid with coral. Its area is 1702 sq. m.

Cedros Island: This sparsely inhabited island on the Pacific coast of Lower California, NW Mexico, has an area of 134 sq. m.

Biscayne Bay (Islandia) National Park.
(National Park Service photograph by
M. W. Williams)

Chandeleur Islands: A popular island group for fur trappers and fishermen
in the Gulf of Mexico, SE Louisiana.

Charlton Island: A terminal for oceangoing ships, this 113 sq. m. (19 m.
long and 9 m. wide) island is at the head of James Bay, Northwest
Territories.

Chauques Islands: A sixteen-island archipelago off the E coast of Chiloé
Island, S Chile.

Chaves Island: This volcanic island is also called Santa Cruz or Indefatigable
Island. It is one of the Central Galápagos Islands, Ecuador, in the
Pacific. It is 389 sq. m. in area.

Cheju Island: Site of an extinct volcano, this Korean island lies between the
Cheju Strait and the East China Sea. It has an area of 713 sq. m.

Chios (Khios): Famous for the Homeridae school of epic poets in classical
times, this island is one of the places which claim to be the birthplace
of Homer. It is in the Aegean, Greece, off Asiatic Turkey and its area is
312 sq. m.

Chonos Archipelago: Running 130 m. N–S, this archipelago consists of over
1000 islands in the Pacific off S Chile. They are uninhabited with the
exception of a few Indians.

Chuginadak Island: The site of a 5680-foot active volcano. Mt. Cleveland,
this is the largest of the Islands of Four Mountains, Aleutian Islands,
SW Alaska. It is 14 m. long and 3–8 m. wide.

Chusan Archipelago: Noted for its strong tidal currents and frequent fogs,
Chusan, the major island of this group in the East China Sea, is 22 m.
long and 10 m. wide.

Cobourg Island (Coburg): The breeding ground for large numbers of murres,
this 22 m. long and 4–14 m. wide island is in Baffin Bay at the entrance
to Jones Sound, Northwest Territories.

Coco, Cayo: A low and swampy coral island off E Cuba in the Old Bahama
Channel, it is part of the Camagüey Archipelago, is 25 m. long and up
to 8 m. wide.

329

Thunderheads tower above sea oats and palmetto on Santa Rosa Island in Gulf Islands National Seashore (proposed). (National Park Service photograph by M. W. Williams)

Cod Island (Ogualik): Its entire surface covered by the Kaumajet Mountains, this NE Labrador island is 11 m. long and 10 m. wide.

Colorados, Los: Also called Guaniguanico or Santa Isabel, this archipelago extends 140 m. along the coast of NW Cuba from Cape San Antonio.

Cornwallis Island: This island has a weather station maintained jointly by the U.S. and Canada on its SE coast. It is 2592 sq. m. in area and is one of the Parry Islands.

Cres Island: This 150 sq. m. island has a length of about 40 m. It is in the Adriatic Sea, Yugoslavia.

Croker Island: This island is an aboriginal reservation. It is 126 sq. m. in area with a length of 27 m. and a width of 10 m. It is in the Arafura Sea, off the NE shore of Cobourg Peninsula, Australia.

Crown Prince Frederick Island: 22 m. long and 6 m. wide, this island is off NW Baffin Island in the Gulf of Boothia, Northwest Territories.

Culion Island: A government leper colony and one of the Calamian Islands, Philippines, this 153 sq. m. island is between Palawan and Mindoro.

Cumberland Islands: These form a 60 m. chain of thirty-six rocky islands and scattered coral islets in the Coral Sea between the Great Barrier Reef and Repulse Bay.

Dahlak Archipelago: This flat, desert and largely uninhabited island group in the Red Sea off the coast of Eritrea consists of two main islands— Dahlak (c. 290 sq. m.) and Narah (c. 50 sq. m.)—and 124 smaller islands.

Damar Islands (Danmar, Dammer): The total area of this S Moluccas, Indonesia, island group is 122 sq. m. This group is in the Banda Sea and consists of the volcanic main island Damar and several islets.

Dawson Island: Primarily a sheep-raising island, it has an area 55 m. long and 20 m. wide. Its location is the central Tierra del Fuego Archipelago in Chile.

D'Entrecasteaux Islands: This is a 1200 sq. m. volcanic group of islands in the SW Pacific, 25 m. SE of New Guinea.

Desolation Island: A bleak uninhabited 80 m. long and 10 m. wide island in the Tierra del Fuego Archipelago, Chile.

Devon Island: Having an ice-covered plateau in the east, this 21,606 sq. m. island is situated in the Arctic Ocean, Northwest Territories.

Disko: First reached by Eric the Red in the years between 982 and 985, this island is situated in the Davis Strait just off W Greenland. It has a total area of 3132 sq. m.

Djerba: A fertile and beautiful island of 197 sq. m. area in the Mediterranean Sea just off the S coast of Tunisia.

Dominica: The largest of the West Indian Windward Islands, it is 29 m. long and 16 m. wide.

Douglas Island: The site of the famous Treadwell mine, this 18 m. long and 3–7 m. wide island is located in SE Alaska, between the mainland and Admiralty Island.

Edge Island: The interior of this island is uncharted and a large icefield extends along its SE coast. One of the Spitsbergen group in the Barents Sea of the Arctic Ocean, it has an area of 1942 sq. m.

Efate: Formerly called Sandwich Island, this is the most important island of the New Hebrides in the SW Pacific. It is 50 m. long and 20 m. wide.

Elcho Island: An aboriginal reservation, this island is in the Arafura Sea, off Napier Peninsula of Arnhem Land, Australia. It is 30 m. in length and has a width of 7 m.

Elephant Island: A 25 m. long and 13 m. wide island, it is off Antarctica's Palmer Peninsula and is one of the E South Shetland Islands.

Eleuthera Island: 164 sq. m. in area, this is one of the central Bahama Islands between Great Abaco Island and Cat Island.

Ellef Ringnes Island: One of the Sverdrup Islands in the Arctic Ocean, Northwest Territories, it has a total area of 4266 sq. m.

Ellis Island: A former U.S. immigration station. Its location is Upper New York Bay, SE New York.

Emerald Island: Approximately 20 m. long and 5–10 m. wide, this island is situated in the Ballantyne Strait of the Arctic Ocean, Northwest Territories.

English Company's Islands: Consisting of four rocky islands and numerous islets, this group forms a 50 m. chain in the Arafura Sea parallel with the NE coast of Australia's Arnhem Land.

Eriskay: The Young Pretender, Prince Charles Edward, first landed on Scottish soil here in 1745. The island is in Scotland's Outer Hebrides and is 3 m. long and 1½ m. wide.

Euboea: This, the largest Greek island in the Aegean Sea, is 1457 sq. m. in area. In ancient times its chief cities were Chalcis and Eretria.

Faitsilong Archipelago: This is a group of islands and islets in the Gulf of Tonkin, North Vietnam. It is 20 m. long and 10 m. wide.

Fakarava: Formerly called Wittgenstein Island, this atoll of the Tuamotu Islands is the second largest of the group in the South Pacific. Its lagoon is 32 m. long and 10 m. wide.

Falster: A very fertile island, its location is in the Baltic Sea, separated from S Zealand (Denmark) by Storstrom Strait. It has an area of 198 sq. m.

Farallon Islands: Sometimes called the Farallones, these two groups of islets are waterless, rocky areas used as bird refuges. They form part of San Francisco city and county. Only Southeast Farallon is inhabited.

Farasan Islands (Farsan, Farisan): Belonging to Saudi Arabia, this island group consists of two large and several smaller islands, the largest of which is 35 m. long. It is a long sandy archipelago off the Asiatic coast in the Red Sea.

Fergusson Island: Of volcanic origin, this 518 sq. m. island, the largest of the D'Entrecasteaux Islands, has hot springs and is 30 m. SE of New Guinea.

Findlay Islands: Lougheed Island is the largest of this group of four islands in the Arctic Ocean, Northwest Territories.

Fire Island: Often called the Great South Beach, this narrow barrier beach is in SE New York off the S shore of Long Island. It is roughly 30 m. long.

Fletcher Islands: This group is comprised of McNamara and Dustin Islands. Its location is off Eights Coast, Antarctica, in the Bellingshausen Sea.

332

Flinders Island: After a period of systematic extermination, the remaining Tasmanians were transported to this island, the largest of the Furneaux Islands, in Bass Strait off the NE coast of Tasmania. It has an area of 802 sq. m.

Flores: A 5511 sq. m. island in the Lesser Sundas, Indonesia, this is one of the most beautiful islands with its deep ravines, gemlike lakes, and a rich variety of flowers. Mountainous, it has a dry climate unlike most Indonesian islands.

Fogo (Fire Island): Having an active volcano (9281 feet), this 184 sq. m. island is in the Leeward group of the Cape Verde Islands in the Atlantic. Its volcano is Cano Peak.

Four Mountains, Islands of: Characterized by strong sea currents and fogs, this is a group of five small, uninhabited islands of the Aleutians, SW Alaska. They are volcanic in origin.

Fox Islands: These islands comprise the easternmost group of the Aleutian Islands, SW Alaska. They extend approximately 300 m. SW from the Alaska Peninsula.

Fro Islands: This is a group of small islands in the North Sea which extend in a 20 m. chain off the coast of Norway. The islands are 20 m. from the coast and run parallel to it.

Fuerteventura (Forteventura): Habitually suffering from drought, this second largest of the Canary Islands has a total area of 666 sq. m.

Fyn (Feyn): This is Denmark's second largest island. It is located between S Jutland and Zealand and has an area of 1149 sq. m.

Geographical Society Island: A 55 m. long and 21 m. wide island in the King Oscar Archipelago, E Greenland, in the Greenland Sea.

George Land: Formerly called Prince George Land, this 80 m. long and 15–30 m. wide island is W of the British Channel in W Franz Josef Land.

Goodenough Island: Formerly Morata, this 20 m. long and 15 m. wide volcanic island is one of the D'Entrecasteaux Islands, Territory of Papua, SW Pacific.

Goto Retto (Goto Islands): This 249 sq. m. total land area is a 60 m. chain of islands in the East China Sea, Japan. The largest of the group is Fukae-shima.

Graham Bell Island: The easternmost island of Franz Josef Land in the Arctic Ocean, it is 45 m. long and 25 m. wide.

Grand Bahama Island: Sometimes called Great Bahama Island, this principal island in the Bahama group is in the NW part.

Grand Canary: One of Spain's Canary Islands, it is the most important of the group and its 592 sq. m. attracts numerous visitors. It forms part of Las Palmas Province.

Grande-Comore Island: This is the largest and westernmost of the Comoro Islands in the Mozambique Channel of the Indian Ocean. It has an area of 442 sq. m.

Grande-Terre: This island with a limestone formation forms the eastern half of Guadeloupe with its 219 sq. m.

Grand Manan Island: A popular resort in the Bay of Fundy, SW New Brunswick, it has a 16 m. length and a 7 m. width.

333

Gravina Islands: These islands are part of SE Alaska's Alexander Archipelago. The largest islands of the group are Gravina, Annette, and Duke.

Great Abaco Island: Adjoined NW by Little Abaco Island, this island is located in the N Bahama Islands and is 100 m. long and 14 m. wide.

Great Barrier Island: This island of volcanic origin forms the breakwater for Hauraki Gulf. It has an area of 110 sq. m. and is 55 m. NE of Auckland, New Zealand.

Grenada: A 120 sq. m. island and the southernmost of the West Indian Windward Islands, its capital, St. George's, is at the southwestern corner of the island.

Groote Eylante: An aboriginal reservation and the largest island in the Gulf of Carpentaria, it has an area of 950 sq. m. and is 25 m. off the coast of Australia's Northern Territory.

Guadalcanal: A 2500 sq. m. island, Solomon Islands, SW Pacific, it is mountainous and rises to a height of 8000 feet in Mt. Popomansiu. The Solomon Islanders are mostly Melanesians with some Polynesians on several of the islands.

Guadalupe Island: 150 m. off the coast of Lower California, this 102 sq. m. island is situated in the Pacific, NW Mexico.

Guam: This is the largest and southernmost of the Marianas Islands in the W Pacific and has an area of 216 sq. m. In 1950, Guam was given statutory powers of self-government and Guamanians are now United States citizens.

Halmahera (Hamaheira, Halmahaira): This, the largest of the Moluccas, Indonesia, consists of four peninsulas separated by three bays. It is 6870 sq. m. in area and is also called Jailolo or Gilolo.

Harris (Isle of Harris): This is the S part of the Lewis with Harris Island, Outer Hebrides, Scotland. It is 10 m. wide and 13 m. long and is connected to the larger part by a 1 m. wide isthmus. It is famous for its tweeds.

Hatteras Island: In E North Carolina, its name is sometimes given to the 40 m. section of the Outer Banks lying between Pamlico Sound and the Atlantic.

Heard Island: This island, largely covered by snow and glaciers, is a sub-Antarctic island in the S Indian Ocean. It is volcanic and rises to 11,000 feet at Big Ben Peak. It is 25 m. long and 10 m. wide.

Hermit Islands: This group consists of five islands on a reef 10 m. wide and 12 m. long. It is a coral group in the Bismarck Archipelago, Territory of New Guinea.

Hierro (Ferro): In ancient times, this island was believed to be the spot where the earth ended. With an area of 107 sq. m., it is the smallest and westernmost of the Canary Islands.

Hiiumaa (Khiuma): Characterized by poor, sandy soil, this island is the second largest of Estonia in the Baltic Sea. It has an area of 373 sq. m.

Hitra: Formerly called Hitteren, this island in the North Sea, central Norway, is separated from the mainland by Trondheim Channel. It has an area of 218 sq. m.

Hiva Oa (Hivaoa): Having a 60 m. circumference, this is the second largest of the Marquesas Islands in the South Pacific. It is volcanic in origin.

Hokkaido (Hokushu): Rugged and bleak, this is the northernmost and second largest of Japan's four main islands. It has an area of some 29,600 sq. m.

Holy Island (Lindisfarne): This island was the site of the first establishment of Celtic Christianity in England. It was here that the Lindisfarne gospels were written. An island in the North Sea, it is just off the E coast of Northumberland, England.

Honshu: Two large mountain chains run the length of this island. It is the largest and most important island of Japan (88,000 sq. m.).

Houtman Abrolhos: This archipelago in the Indian Ocean is 35 m. off the coast of western Australia, extends for 50 m. N–S and comprises three coral groups: Wallabi Islands, Easter Islands, and Pelsart Islands.

Hvar Island: This 129 sq. m. island is in the Adriatic Sea, S. Croatia, Yugoslavia, and is Dalmatia's number one island vacation spot.

Hydra: This Greek island in the Aegean Sea played an important role in the Greek war for independence. It is now a society island playing host to

The village of St. James reaches around the harbor at Beaver Island, tranquil Michigan vacation hideaway in northeastern Lake Michigan. The island's natural greenery and its Irish settlers won it the nickname "Emerald Isle." (Courtesy Michigan Tourist Council)

numerous celebrities in the arts and letters as well as other visitors. It has an area of 20 sq. m.

Icaria (Ikaria): This island was named for the Greek mystic hero Icarus, who supposedly fell into the sea near this site. It is a 99 sq. m. island in the Aegean Sea, Greece.

Imbros: Turkey's largest island, its location is in the Aegean Sea off the coast of Gàllipoli Peninsula. It has an area of 108 sq. m.

Inagua: Comprised of two islands, Great Inagua and Little Inagua, this small group is the southernmost of the Bahama Islands. Together, they total 560 sq. m. Great Inagua has an abundance of birdlife.

Investigator Islands: These comprise a 40 m. long chain of islands in the Great Australian Bight, 4 m. off the W coast of Eyre Peninsula. This group consists of Flinders Island, Pearson Island, and Waldegrave Island.

Ionian Islands: The largest islands in this group are Corfu, Leukas, Cephalonia, and Zante. Its location is the Ionian Sea, off the W coast of Greece. Lush Corfu, with its long and varied history, is perhaps the most interesting of these islands and the inspiration for Shakespeare's *The Tempest.*

Iriomote-jima: This is a 144 sq. m. island of volcanic origin. It is part of the Sakishima group in the Ryukyus between the East China and Philippine Seas.

Isabela Island (Albemarle): This 2249 sq. m. island is the largest of the Galápagos Islands in the Pacific. Like all the other islands in this group, it is volcanic in origin.

Islay: This is an island of the Inner Hebrides, Scotland, and the most southerly of the Hebrides. In area it is 234 sq. m. including Oversay, a small island off its SW end.

335

ISLANDS

Iturup Island: This is the largest and most important of the Kurile Islands, Russian SFSR. It has an area of 2587 sq. m. and is one of the islands reputed to have belonged to Japan since antiquity.

Iviza (Ibitza): Like the other Balearic Islands, this one is chiefly agricultural with sheep and pig raising as sidelines. It is the smallest of the chief Balearic Islands in the W Mediterranean. Its area totals 221 sq. m. with adjacent isles.

Iwo Jima: This 8 sq. m. island is the largest and most important of the W Pacific Volcano Islands. The site of Japanese air bases in World War II, it was occupied by U.S. forces in March, 1945, after heavy fighting.

Japen Islands: Also spelled Jappen, Yapen and Yappen, this island group in Geelvink Bay belongs to Netherlands New Guinea. In all, the area comprises 10,907 sq. m. The group consists of three islands, the largest of which is Japen (100 m. long and 15 m. wide).

Jardines de la Reina: Comprising more than 400 keys, this archipelago of coral reefs off the Caribbean coast of E Cuba is approximately 85 m. long NW–SE.

Jenny Lind Island (Lind): This 17 m. long and 10 m. wide island is in Queen Maud Gulf off SE Victoria Island, Northwest Territories.

Jersey: This 45 sq. m. island is the largest and southernmost of the Channel Islands. It is also the home of the famous breed of cattle of the same name.

Jolo Island (Sulu): This is the chief island and capital of the Sulu province. It is part of the Sulu Archipelago, Philippines, between the Sulu and Celebes Seas and has an area of 345 sq. m.

Kanaga Island: Rising to 4416 feet at Kanaga Volcano, this 30 m. long and 4–8 m. wide island is one of the Andreanof Islands, Aleutian Islands, SW Alaska.

Kangaroo Island: Structurally a part of Mount Lofty Ranges on the mainland of South Australia, this island lies 27 m. S of Yorke Peninsula in the Indian Ocean. It is 90 m. long and 33 m. wide.

Kangean Islands: This group has a total area of 258 sq. m. and is comprised of three islands surrounded by c. sixty islets. The largest island is Kangean (188 sq. m.) and the group is in the Java Sea, Indonesia.

Karimata Islands: This island group in the Karimata Strait, Indonesia, off the SW coast of Borneo comprises two large islands, Karimata and Serutu (Seroetoe), and around sixty islets.

Karimun Islands: Also spelled Karimoen, this island group is part of the Riouw Archipelago, Indonesia, in the South China Sea. The largest island in the group, Kundur (Koendoer), is 18 m. long and 10 m. wide.

Karkar: Formerly Dampier Island, this island of volcanic origin still has an active volcano. It is 140 sq. m. in area and its location is the Territory of New Guinea in the SW Pacific, 9 m. NE of New Guinea.

Kauai: 551 sq. m. in area, this island is the fourth largest of the Hawaiian Islands and geologically the oldest of the group.

Kebao Island: A triangular island in the Gulf of Tonkin, N Vietnam, it has a length of 15 m. and a width of 10 m.

King Christian Island: This island is situated in the Maclean Strait off Ellef Ringnes Island, Northwest Territories. It is 9 m. wide and 17 m. long.

King George Islands: This island group consists of fifteen small islands in Hudson Bay, Northwest Territories. It covers an area 30 m. long and 20 m. wide.

King Oscar Archipelago: This group of islands in the Greenland Sea, E Greenland, is between Franz Josef Fjord and King Oscar Fjord.

Kiska Island: One of SW Alaska's Rat Islands, Aleutian Islands, this island is 20 m. long and 2–7 m. wide.

Komsomolets Island: 65 percent of this island's 3570 sq. m. is covered with glaciers. It is the northernmost island of the Severnaya Zemlya archipelago in the Arctic Ocean.

Kong Karls Land: Also referred to as King Charles Islands or King Karl Islands, this group consists of three islands and several islets in the Barents Sea of the Arctic Ocean. In area 128 sq. m., it is part of Svalbard (Norwegian possession).

Korcula Island (Korchula): A 107 sq. m. Dalmatian island in the Adriatic, S Croatia, Yugoslavia.

Korean Archipelago: This is the name sometimes given to island groups off the SW coast of Korea.

Kos: Hippocrates was born on this second largest island of the Dodecanese, Greece, in the Aegean Sea.

Kosciusko Island: This 25 m. long and 5–12 wide island is part of SE Alaska's Alexander Archipelago.

Krk Island: Formerly a part of Istria, this largest of the Yugoslav islands in the N Adriatic, NW Croatia, is 165 sq. m. in area and, for the most part, barren.

Kunashir Island: This volcanic island is the southernmost and second largest of the main Kurile Islands chain, Russian SFSR. In area 1548 sq. m., it is one of three islands reputed to have belonged to Japan since ancient times.

Kupreanof Island: This island is located in S Alaska between Raspberry Island and Kodiak Island. It is 52 m. long and 20–30 m. wide.

Kwajalein: This largest atoll of the Marshall Islands in the W central Pacific is 6 sq. m. in area. It consists of ninety-seven islets on a lagoon and was a Japanese base in World War II before it was captured by the United States in 1944.

Kythera (Kithira): This Mediterranean island is 108 sq. m. in area and its location is at the mouth of the Gulf of Laconia, Greece.

Kyushu (Kiushiu, Kyusyu): This rugged and bleak island belongs to Japan and has a total area of 13,770 sq. m. Its location is just S of Honshu between the Philippine and East China Seas.

Lanai: This island has an area of 141 sq. m. and is volcanic in origin. Although mountainous, it is one of eight inhabited islands in the Territory of Hawaii about 7 m. W of Maui.

Lands Lokk: Located in the Arctic Ocean, Northwest Territories, off NW Ellesmere Island, this island is 8 m. wide and 20 m. long.

Langoy: The site of important fisheries, this 332 sq. m. island is part of the Vesteralen group in the North Sea, N Norway.

Lanzarote: One of the volcanic Canary Islands, Las Palmas Province, Spain, this island has an area of 307 sq. m.

Laurie Island: This long and narrow island has a length of 110 m. and a width of 1–3 m. It is one of the South Orkney Islands in the S Atlantic.

Leukas (Levkas): Sometimes identified with ancient Ithaca, this 114 sq. m. Greek island is one of the Ionian Island group in the Ionian Sea.

Lewis with Harris: Rich with relics of a history of lost causes and stubborn though hopeless loyalties is this island, the largest and northernmost of Scotland's Outer Hebrides. It has an area of 825 sq. m.

Leyte: This mountainous island, which rises to a height of 4425 feet, is one of the Visayan Islands, E central Philippines, between Luzon and Mindanao. It has a total area of 2785 sq. m.

Little Andaman Island: This southernmost of the Andaman Islands in the Bay of Bengal is separated from the main group by the Duncan Passage. It is 26 m. long and 16 m. wide.

338 *Livingston Island:* One of the South Shetland Islands, off Palmer Peninsula, Antarctica, this island is 32½ m. long and 4–16 m. wide.

Lomblen (Kawula, Kawoela): The largest of the Solar Islands, Lesser Sundas, Indonesia, between the Flores and Savu Seas, this irregularly shaped island has an area of 499 sq. m.

Londonderry Island: A 27 m. long island in the Pacific's Tierra del Fuego, Chile.

Lougheed Island: This is the largest of the Northwest Territories' Findlay Islands in the Arctic Ocean. It is 504 sq. m. in area.

Louisiade Archipelago: This island group includes approximately ten volcanic islands and several coral reefs. Its location is the Territory of Papua, SW Pacific, 125 m. SE of New Guinea.

Luzon: This island houses Manila, the capital and principal port of the Philippine Islands. It is also the largest of these islands with an area of 40,420 sq. m. It is at the W end of the archipelago.

Lyakhov Islands: This is the S group of the New Siberian Islands between the East Siberian and Laptev Seas. Their total area is 2660 sq. m.

Madame Island: Sometimes referred to as Isle Madame, this small island has an area of 108 sq. m. (12 m. long and 9 m. wide) and lies S of Cape Breton Island in E Nova Scotia.

Madura: Pamekasan is the principal town of this Indonesian island in the Jave Sea just off the NE coast of Java. Madura's total area is 1762 sq. m.

Northern Service officer shows examples of handicrafts produced by other Eastern Arctic Eskimos to a group of children at Pangnirtung, Baffin Island, Canada. (Courtesy National Film Board of Canada)

Mafia Island: Identified with the island of Menuthias, which was mentioned in the oldest sailing guide to the East African coast and was visited by tourists from the Mediterranean 1400 years before the discovery of America, this 170 sq. m. island is in the Indian Ocean, E Tanganyika.

Magdalen Islands: Their French name, "Iles de la Madeleine," is often used in reference to this island group in E Quebec's Gulf of St. Lawrence. It consists of nine main islands and numerous islets for a total area of 102 sq. m.

Mageroy: On this island's N coast is Knivskjelloden, the northernmost point of Europe. In area 106 sq. m., this island is situated in the Barents Sea of the Arctic Ocean, N Norway.

Mainland: Identified with Ultima Thule, this island has an area of 406½ sq. m. and is the largest of the Shetlands, Scotland.

Majorca: This largest of the Balearic Islands, Spain, in the Mediterranean has an area of 1405 sq. m. Beautiful Majorca is a typical example of limestone scenery with craggy peaks and fertile valleys.

Malaita: On this island, the coastal people barter fish for the vegetables which the bush-dwellers grow. Volcanic in origin, this island is the most populous of the Solomon group, SW Pacific. It has an area of 1572 sq. m.

Malay Archipelago: This name is used for the island group between the Malay Peninsula and Indochina (NW) and Australia and New Guinea (SE). It includes the islands of Indonesia and the Philippines and sometimes New Guinea. It also includes the East Indies.

Maltese Islands: Generally called Malta, this 122 sq. m. archipelago in the central Mediterranean is comprised of Malta, Gozo, and several smaller islands.

339

ISLANDS

Mansel Island: This 62 m. long and 4–30 m. wide island has been a reindeer reserve since 1920. Its location is off N Ungava Peninsula, in the Hudson Bay, Northwest Territories.

Manus (Admiralty Island): This largest of the Admiralty Islands, Bismarck Archipelago, Territory of New Guinea, has an area of 633 sq. m. This island became widely known due to the anthropological studies of Margaret Mead. This island's chief pursuits are coconut growing and pearl fishing.

Margarita Island: This is actually two islands joined by a sandspit. It has a total area of 444 sq. m. and is horseshoe-shaped. Its location is the Caribbean, NE Venezuela.

Marion Island: Annexed by South Africa in 1947, this 13 m. long and 8 m. wide sub-Antarctica island is just SW of Prince Edward Island in the S Indian Ocean.

Marsh Island: A low marshy island between the Gulf of Mexico, Vermilion and West Cote Blanche Bays, S Louisiana, this island is 21 m. long and 2–10 m. wide.

Masbate: One of the chief gold-bearing areas of the Philippines and one of the Visayan group, it has an area of 1262 sq. m.

Mascarene Islands: This island group in the Indian Ocean consists of Réunion, Mauritius, and several others.

Matagorda Island: This low, sandy island between the Gulf of Mexico and the San Antonio and Espiritu Santo Bays, S Texas, is 36 m. long and 1–4 m. wide.

Maui: One of the inhabited Hawaiian Islands, Maui is the second largest with 728 sq. m. It is separated from the island of Hawaii by the Alenuihaha Channel.

Melanesia: Meaning "black islands," this is the name of one of the three main divisions of the Pacific islands. It includes the Fiji Islands, New Caledonia, Loyalty Islands, New Hebrides, Solomon and Santa Cruz Islands, Admiralty Islands, Louisiade and Bismarck Archipelagos, and D'Entrecasteaux Islands.

Melbourne Island: East of the base of Kent Peninsula in Queen Maud Gulf, Northwest Territories, is this 18 m. long and 10 m. wide island.

Melville Island: This is the largest of the Parry Islands in the Arctic Ocean, Northwest Territories, and it has an area of 16,503 sq. m.

Mentawai Islands (Mentawei): Consisting of about seventy islands, the largest of which is Siberut, this volcanic group has a total area of 2354 sq. m. and is located off the W coast of Sumatra, Indonesia, in the Indian Ocean.

Merrit Island: This island is situated in E Florida between the Banana River and Indian River lagoons. It is 30 m. long and up to 7 m. wide.

Micronesia: One of the three main divisions of the Pacific islands, it consists of four archipelagos: Caroline, Marshall, Marianas, and Gilbert. The Micronesian people are a mixture of Melanesian, Polynesian, and Malay stock rather than a distinct racial group.

340

Midway: The Battle of Midway (June 3–6, 1942) was one of the decisive battles in World War II. It includes a 2 m. sq. atoll and two islets halfway across the N Pacific.

Mindanao: A mountainous and the second largest island of the Philippines, it is 36,537 sq. m. in area and lies at the southern end of the group.

Mindoro: This 3759 sq. m. island in the Philippines is between the Mindoro and Tablas Straits.

Minorca: This island boasts one of the finest harbors in the Mediterranean. It is the second largest of Spain's Balearic Islands and has a total area of 271 sq. m.

Molokai: This volcanic and mountainous island is one of the inhabited Hawaiian Islands in the State of Hawaii. It has an area of 259 sq. m.

Mona: This is the Roman name for an island sometimes identified with the Isle of Man, sometimes with Anglesey.

Monte Bello Islands: Surrounded by coral reefs, the largest island in this coral group is Barrow Island. Its location is the Indian Ocean off the NW coast of western Australia.

Moresby Island: This is one of the Queen Charlotte Islands, W British Columbia in the Pacific. It has a total of 1060 sq. m.

Muckle Flugga: At the site of the North Unst Light, this northernmost of the Shetland Islands is just N of Unst.

Mussau (St. Matthias Islands): This island group is comprised of two volcanic islands—Mussau (c. 160 sq. m.) and Emireau (20 sq. m.)— and several coral islets. Its location is the Bismarck Archipelago, Territory of New Guinea, SW Pacific.

341

Namhae Island: Almost divided in two, this island has three peninsulas. It is 115 sq. m. in area and is situated in Korea's Cheju Strait, which is almost joined to the southern coast of the mainland.

Natuna Islands (Natoena): This island group of Indonesia, between Borneo and the Malay Peninsula in the South China Sea, is comprised of Great Natuna (the largest) and two small groups: North Natuna Islands and South Natuna Islands. These islands are mostly low and wooded.

Near Islands: This is the westernmost group of the Aleutian Islands, SW Alaska. It is uninhabited with the exception of Attu, the largest island.

Negros: Its chief products are rice and sugarcane and it is the fourth largest of the Philippine Islands in the Visayan Islands, between Panay and Cebu Islands with a total area of 4905 sq. m.

New Georgia: A volcanic island group of the Solomon Islands, SW Pacific, it has a total area of 2000 sq. m. The largest island of the group is New Georgia (20 m. wide and 50 m. long).

New Siberian Islands: This Russian SFSR archipelago between the Laptev and East Siberian Seas is characterized by few inhabitants, sparse tundra, and mammoth fossils. Its area is 11,000 sq. m.

New Zealand: A dominion of the British Commonwealth of Nations in the S Pacific, it consists of North and South Islands with an area of 103,416 sq. m.

ISLANDS

Nootka Island: First visited in 1774 by the Spanish explorer Juan Perez, the island was later examined in detail by Capt. James Cooke on his third voyage of discovery. This 206 sq. m. island is in SW British Columbia off the W coast of Vancouver Island.

Normanby Island: This 400 sq. m. island is one of the D'Entrecasteaux Islands and is volcanic in origin. It is part of the Territory of Papua, SW Pacific.

Northeast Land: An almost completely glaciated island of the Spitsbergen group in the Barents Sea, it is 5710 sq. m. in area.

North Island: Although it is the smaller of New Zealand's two main islands with an area of 44,281 sq. m. it is the more populous. It has several mountain ranges and an active and a dormant volcano.

North Uist: One of the islands of the Outer Hebrides, Scotland, it has a total area of 118 sq. m.

Nunarssuit: Formely called Desolation Island, this 20 m. long and 3–9 m. wide island is off SW Greenland, off the SW side of the mouth of Kobbermine Bay.

Nunivak Island: Treeless and characterized by heavy fogs, this W Alaska island in the Bering Sea is separated from the mainland and Nelson Island by the Etolin Strait. It is 56 m. long and 40 m. wide.

Oahu: 589 sq. m. in area, this is the third largest of the Hawaiian Islands. It is an inhabited island and volcanic in origin.

Obi Islands (Ombi Islands): This island group in the N Moluccas, Indonesia, in the Ceram Sea consists of Obir (or Obira), which is 52 m. long and 28 m. wide, and small offshore islands.

Okinawa: This is the largest of the Okinawa Islands with a total area of 467 sq. m. This island was the scene of a decisive and bitter battle in World War II.

Okinawa Islands: Comprising 579 sq. m. and stretching in a 70 m. chain, this central group of the Ryukyu Islands is between the East China and Philippine Seas.

Oktyabrskaya Revolyutsiya Island (October Revolution): Almost half-covered with glaciers, this central island of the Severnaya Zemlya Archipelago in the Arctic Ocean is 5400 sq. m. in area.

Ontong Java: Also called Lord Howe Island, this atoll of the Solomon Islands, SW Pacific, is comprised of four islands on a reef 30 m. long and 20 m. wide.

Ostero: 111 sq. m. in area, this is the second largest of the Faeroe Islands, a group of twenty-one volcanic islands under Danish sovereignty.

Ottawa Islands: This group of twenty-four small islands in Hudson Bay, off NW Ungava Peninsula, covers an area 70 m. long and 50 m. wide.

Outer Banks (The Banks): This chain of sandy barrier islands extends along the North Carolina coast.

Padre Island: A long narrow barrier island in S Texas between Laguna Madre and the Gulf of Mexico, it is about 115 m. long and generally under 2 m. wide.

Cliffs of the east shore of Stockton Island, Apostle Islands, Wisconsin. (National Park Service photograph by M. W. Williams)

Pag: Separated from the mainland by the Velebit Channel, this island in W Croatia, Yugoslavia, is in the Adriatic and is 114 sq. m. in area. It is famed for its lace.

Pagi Islands (Pagai): A group of the Mentawai Islands, Indonesia, off the W coast of Sumatra, it consists of North Pagi (25 m. long and 17 m. wide), South Pagi (42 m. long and 12 m. wide), and many islets.

Palawan: Formerly called Paragua, this westernmost of the large islands in the Philippines, between the Sulu and South China Seas, is 4550 sq. m. in area.

Palmer Archipelago: This island group of Antarctica is off the NW coast of Palmer Peninsula, SW of the South Shetlands.

Palm Island: Comprised of approximately twenty islands and rocks, this coral group lies in the Coral Sea within the Great Barrier Reef off the E coast of Queensland, Australia.

Palmyra: This atoll in the central Pacific is comprised of fifty-five islets in the Line Islands.

Patmos: According to the Christian tradition, St. John wrote the Biblical book of Revelation on this 13 sq. m. island in the Greek Dodecanese.

Pavlof Islands: Consisting of seven islands, the largest of which is Dolgoi Island (10 m. long and 7 m. wide), this island group is situated at the entrance to Pavlof Bay, off the SW Alaska Peninsula.

Pearl Islands: Named for their important pearl fisheries, this group of 183 islands, thirty-nine of them fairly large, lies in the Gulf of Panama about 40 m. SE of Panama.

Pemba: Noted for its clove production, this 380 sq. m. coral island (including offshore islets) is part of the Zanzibar protectorate. Its location is in the Indian Ocean off the E coast of Africa, and it is separated from the Tanganyikan coast by 40 m. wide Pemba Channel.

Peter I Island: Named for the founder of the Russian Navy, this island is located in the Bellingshausen Sea off the coast of Antarctica. In area 100 sq. m., it has been a dependency of Norway since 1933.

343

Philippine Islands: An archipelago of 7083 islands in the Pacific SE of China, it was formerly under the guardianship of the U.S. and is now an independent republic.

Phuket: Known for its large tin deposits, this 206 sq. m. island is in the Andaman Sea, S Thailand, off the W coast of the Malay Peninsula.

Pico Island: This 167 sq. m. island is in fact the summit of a mountain— a still active volcano over 4 m. high. One of the Azores, it is in the central part of the group in the Atlantic Ocean.

Pines, Isle of: Off SW Cuba, to which it belongs, this island has a total area of 1182 sq. m.

Pitt Island: Separated from the mainland by Grenville Channel and by Principe Channel from Banks Island, this island in Hecate Strait, W British Columbia, has an area of 528 sq. m.

Polillo Islands: The largest of this Philippine group in the Philippine Sea off the E coast of Luzon, is Polillo Island (234 sq. m.). The total area for this group is approximately 295 sq. m.

Polynesia: The name, meaning "many islands," is one of the three main divisions of the Pacific islands. It includes the islands of Hawaii, Samoa, Tonga, Tokelau, Society, Marquesas, Cook, Ellice and Easter.

Pomona (Mainland): The largest of Scotland's Orkney Islands, it is 17 m. wide and 25 m. long.

344 *Possession Islands:* A group of nine islands in the Ross Sea, its location is off the NE coast of Victoria Land, Antarctica.

Prince of Wales Island: Although it has many namesakes, the "real" Prince of Wales Island is located in SE Alaska's Alexander Archipelago, of which it is the largest island with an area of 2231 sq. m.

Prince Patrick Island: A weather station maintained jointly by Canada and the United States is situated on this island, one of the Parry Islands in the Arctic Ocean, Northwest Territories. It has an area of 6696 sq. m.

Qishm (Kishm, Qeshm): The largest island in the Persian Gulf, SE Iran, in the Strait of Hormuz, it is 70 m. long and 7–20 m. wide.

Raja Ampat Islands: An island group between the Pacific and the Ceram Sea off Vogelkop Peninsula, Netherlands New Guinea, its chief islands are Waigeu and Misool.

Ralik Chain (Ralick): Sometimes called the Sunset Group, this W group of the Marshalls in the W central Pacific is comprised of three coral islands and fifteen atolls.

Rangiroa: Formerly known as Dean's Island, this is the largest of the Tuamotu Islands, French Oceania. It comprises twenty islets surrounding a lagoon 45 m. long and 15 m. wide.

Ratak Chain: Sometimes called the Sunrise Group, this E group of the Marshall Islands, W central Pacific, is comprised of two coral islands and fourteen atolls.

Rat Islands: Extending for about 110 m. E-W, this group of the Aleutians is W of the Andreanof Islands and SE of the Near Islands.

Recherche Archipelago: Also referred to as the Archipelago of the Recherche, this island group extends 120 m. E-W off the S coast of western Australia in the Indian Ocean.

Resolution Island: Off SE Baffin Island, this 1029 sq. m. island is at the E entrance to Hudson Strait.

Revillagigedo Island: Separated from the mainland by Behm Channel and from Prince of Wales Island by Clarence Strait, this 1120 sq. m. island is part of SE Alaska's Alexander Archipelago.

Riouw Archipelago: Sometimes spelled Riau, it is one of the important Indonesian island groups in the South China Sea. Its location is at the entrance to the Strait of Malacca and its total area is 2279 sq. m. Its largest island is Bintan(g).

Romano, Cayo: This island forms part of the Camagüey Archipelago. Coral in origin, its location is in the Old Bahama Channel off the N coast of Cuba. It is 55 m. long and 10 m. wide.

Rooke Island (Rook): Also called Umboi, this island of volcanic origin is 300 m. in length. It is part of the Bismarck Archipelago, Territory of New Guinea, SW Pacific.

Roosevelt Island: 79 m. long and 35 m. wide, this island comprises the E section of the Ross Shelf Ice, Antarctica.

Ross Island #1: Also known as James Ross Island, this 34 m. long and 27 m. wide island lies in the Weddell Sea, Antarctica, just off the NE tip of Palmer Peninsula.

Ross Island #2: The site of Mt. Erebus and Mt. Terror, this island is 39 m. long and 37½ m. wide. Its location is the outer edge of Ross Shelf Ice in the W part of the Ross Sea, Antarctica.

Roti: This island is part of Indonesia; it is 50 m. long and 14 m. wide and has a total area of 467 sq. m. It is separated from Timor by the 10 m. wide Roti Strait.

Royal Geographical Society Islands: An island group consisting of four small islands and many islets, its location is in the Northwest Territories between the S end of Victoria Strait and the NE side of Queen Maud Gulf.

Saare (Sarema, Saaremaa): This is the largest island of Estonia in the Baltic Sea at the entrance to the Gulf of Riga. It has an area of 1046 sq. m.

Sabinal, Cayo: A 25 m. long and approximately 6 m. wide coral island at the entrance to Old Bahama Channel off NE Cuba.

St. Croix Island: This is the largest of the United States Virgin Islands. It has an area of 82 sq. m.

St. Lucia: Of the larger Windward Islands, this one is perhaps the most beautiful with its famous Pitons rising sheer from the sea. It is 233 sq. m. in area and is part of the West Indies. It is separated from Martinique by St. Lucia Channel and from St. Vincent by St. Vincent Passage.

St. Vincent: One of the central Windward Islands, West Indies, and the site of Mount Soufrière, it is 133 sq. m. in area.

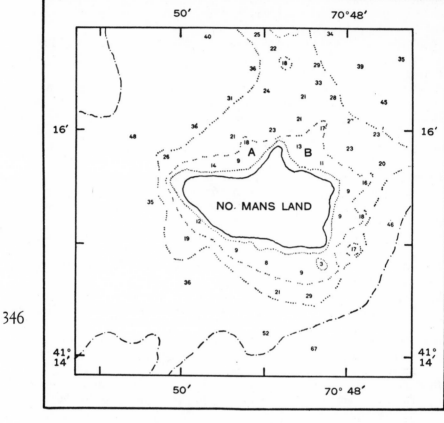

Figure 1. Chart of No Man's Land, Massachusetts. Soundings in fathoms.

Sakishima Islands: A 34 sq. m. S group of the Ryukyu Islands, between the East China and Philippine Seas, it formed a native civil administration in 1951.

Salamis: In the channel between Salamis and the mainland, the place of the defeat of the Persians under Xerxes by the Greeks under Themistocles (480 B.C.), it is 39 sq. m. in area and its location is in the Saronic Gulf of the Aegean Sea, E central Greece.

Salsette Island: The "home" of the famous Kanheri cave-temples, this Indian island is in the Arabian Sea off Bombay. It is 28 m. long and up to 15 m. wide.

Salut, Iles du: This archipelago off the coast of French Guiana comprises Devil's Island, Ile Royale, and St. Joseph Island.

Samar: 5050 sq. m. in area, this island is the third largest of the Philippines in the Visayan group between the Samar and Philippine Seas.

Samos: The birthplace of Pythagoras, this Greek Aegean island off the Mycale Peninsula of Turkey is 194 sq. m. in area.

San Cristóbal Island: Also referred to as Chatham Island, this is one of the E Galápagos Islands, Ecuador, in the Pacific. It is 195 sq. m. in area.

Sandwip Island: The easternmost island of the Ganges Delta, E Pakistan, in the Bay of Bengal, it is 25 m. long and 3–9 m. wide.

San Juan Islands: An archipelago of 172 islands at the N end of Puget Sound, NW Washington.

San Miguel Island: This is the largest of the Pearl Islands, E Panama. in the Gulf of Panama. It is 8 m. wide and 17 m. long.

Santa Catalina Island: Linked to the mainland by a steel bridge at Florianópolis, Brazil, this island is 33 m. long and up to 10 m. wide. It lies in the Atlantic Ocean and forms part of Santa Catarina state.

Santa Cruz Islands: The largest island of this volcanic group of the Solomon Islands is Ndeni. Its location is in the SW Pacific and its total area is 370 sq. m.

Santiago Island: Also known as San Salvador Island and James Island, this 203 sq. m. island is part of the central Galápagos Islands, Ecuador.

São Miguel Island: Also called St. Michael Island, this is the largest of the Azores, Atlantic Ocean, in the E part of the group. It is 288 sq. m. in area.

São Tiago Island: Also called Santiago Island, this 383 sq. m. island is the largest of the Leeward group of the Cape Verde Islands.

347

Savaii: Built up around rugged mountains, this volcanic island of Western Samoa is the largest and most westerly of the group. Formerly called Chatham Island, it is 703 sq. m. in area.

Schouten Islands: Also referred to as the Misore Islands, this archipelago off NW New Guinea at the entrance to Geelvink Bay is 1231 sq. m. in area. It is comprised of Biak, Supiori, Numfor, and several smaller islands.

Serrano Island: Sometimes called Little Wellington Island, this 40 m. long island lies just N of Wellington Island, Aysén Province, S Chile.

Shannon Island: The 7–10 m. wide Shannon Sound separates this island from the mainland of NE Greenland. The island itself is 35 m. long and 5–15 m. wide.

Shantar Islands: The largest island of this group is Bolshoi Shantar Island (35 m. long and 28 m. wide). The group is comprised of four large and eight small islands and lies in the SW section of the Sea of Okhotsk. Its total area is 965 sq. m.

Sherbro Island: Separated from mainland Sierra Leone by Sherbro River and Shebar Strait, this island lies in the Atlantic 65 m. SSE of Freetown, W. Africa. It is 30 m. long and up to 15 m. wide.

Shikoku: Although a rather large island with an area of 6860 sq. m., it is actually the smallest of Japan's main islands. It lies S of Honshu, E of Kyushu between the Inland and Philippine Seas.

ISLANDS

Shortland Islands: The largest of this island group is Shortland Island (10 m. long and 8 m. wide). Of volcanic origin, these islands comprise 200 sq. m. and are part of the SW Pacific Solomon Islands.

Sicily: This largest island in the Mediterranean has 9924 sq. m. and forms a department of Italy. It is separated from the SW tip of the Italian mainland by the Strait of Messina.

Sir Edward Pellew Islands: The largest of this island group is Vanderlin Island (17 m. long, 9 m. wide). It has an area of 800 sq. m. and its location is the Gulf of Carpentaria off the E-coast of Australia's Northern Territory.

Sir Joseph Banks Islands: Consisting of twenty islands, islets, and rocks, this group lies in Spencer Gulf off the SE coast of Eyre Peninsula, South Australia.

Solor Islands: This group of the Lesser Sundas, Indonesia, in the Flores Sea is comprised of Adonara, Solor, and Lomblen.

Solovetski Islands (Solovetskye): The largest island of this group is Solovetski Island (110 sq. m.). Its location is in the White Sea at the entrance to Onega Bay, Russian SFSR.

South Island: This 58,093 sq. m. island's most impressive feature is its range of the Southern Alps. It is the larger of New Zealand's two main islands.

South Uist: A 2810 sq. m. island of the Outer Hebrides, Scotland, between Benbecula and Barra.

Staten Island (I. de los Estados): A 209 sq. m. island in the South Atlantic 18 m. E of the SE tip of the main island of Tierra del Fuego.

Sula Islands (Soela): The largest of this island group is Taliabu. The entire group has an area of 1873 sq. m. It is a group of the N Moluccas, Indonesia, between Celebes and Obi Islands in the Molucca Sea.

Sumba (Soemba): Formerly known as Sandalwood Island, this 4306 sq. m. island is part of Indonesia's Lesser Sundas in the Indian Ocean.

Sunda Islands (Soenda): Divided into Greater and Lesser Sundas, this island group in Indonesia comprises the W part of the Malay Archipelago between the South China Sea and the Indian Ocean.

Sverdrup Islands: An archipelago in the Arctic Ocean, Northwest Territories, it includes Axel Heiberg, Ellef Ringnes, Armund Ringnes, and Meighen Islands as well as several smaller ones.

Tagula: Volcanic in origin, this 50 m. long and 15 m. wide island is the largest in the Louisiade Archipelago, Territory of Papua, in the Pacific.

Talaud Islands (Talaut): Sometimes called Tolaus Islands, sometimes spelled Talaur, this Indonesian group of islands in the Pacific S of Mindanao is 495 sq. m. in area.

Taliabu (Taliaboe): This 68 m. long and 25 m. wide island is the largest of the Sula Islands, N Moluccas, Indonesia, in the Molucca Sea.

Tanega-shima: Between the East China and Philippines Seas lies this Japanese island 20 m. S of Kyushu. It is 126 sq. m. in area.

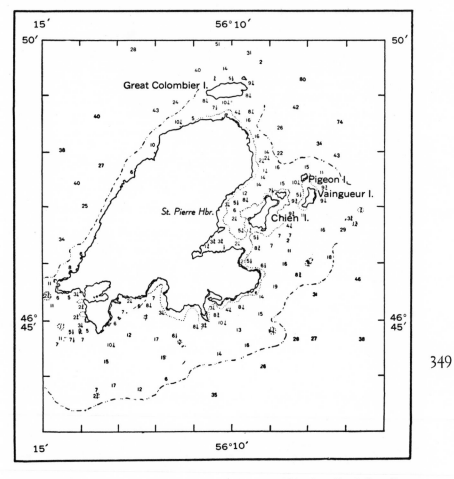

Figure 2. Chart of St. Pierre Island, off the south coast of Newfoundland. Soundings in fathoms.

Tanimbar Islands: Also called Timorlaut or Timorlaoet, and sometimes spelled Tenimbar, this 2172 sq. m. island group of the S Moluccas, Indonesia, lies in the Banda Sea.

Tawitawi Group (Tawi-Tawi): The largest island of this group is Tawitawi (229 sq. m.) although most of the population is on the small offshore islands. It is part of the Sulu Archipelago of the Philippines between the Sulu and Celebes Sea.

Tenerife (Teneriffe): The largest of the Canary Islands, Spain, in the Atlantic, it has an area of 794½ sq. m.

Thasos (Thassos): Famous for its gold mines in ancient times, this 170 sq. m. island lies in the N Aegean Sea, Greece, off the mouth of the Mesta River.

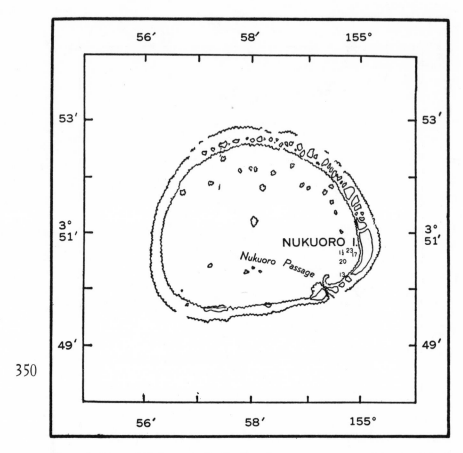

Figure 3. Chart of Nukuoro Atoll, Caroline Group, western tropical Pacific. Soundings in fathoms.

Thousand Islands: A group of about 100 coral islets of Indonesia, they lie in the Java Sea off the NW coast of Java.

Tiburon Island: Located in the Gulf of California off the W coast of Sonora State, Mexico, this island has an area of 458 sq. m.

Togian Islands: Sometimes referred to as the Schildpad Islands, this island group of Indonesia lies in the Gulf of Tomini, N Celebes. The largest island is Batudaka (18 m. long and 8 m. wide).

Tonga: Also called the Friendly Islands, this group in the S Pacific, NE of Sydney, is 250 sq. m. in area.

Tongatabu: Tongatabu is the largest of this coral island group of S Tonga in the South Pacific.

Tres Marias Islands: Also called Las Tres Marias, this archipelago in the Pacific off the coast of Nayarit State, W Mexico, has a total area of 100 sq. m.

Tubuai Islands: Sometimes referred to as Austral Islands, this volcanic group in the South Pacific belongs to French Oceania.

Tukangbesi Islands: Also spelled Toekangbesi, this Indonesian island group lies between the Flores and Molucca Seas. The largest island is Wangi-wangi with 60 sq. m. of area.

Turks Islands: The total area of the Turks and Caicas groups is 202 sq. m. The E sector of these two groups is a dependency of Jamaica, West Indies.

Tuscan Archipelago: Comprised of Elba, Monte Cristo and others, this island group between the Corsica and Tuscany coasts is in the Tyrrhenian Sea, Italy, and has an area of 115 sq. m.

Ulithi (Uluthi): This atoll in the W Caroline Islands, W Pacific, is 19 m. long and 10 m. wide.

Umnak Island: One of the Fox group in SW Alaska's Aleutian Islands, this island is 83 m. long and 2–18 m. long.

Unalaska Island: Just E of Umnak Island is this 30 m. long and 6–30 m. wide island of the Fox group in SW Alaska's Aleutian Islands.

Unimak Island: This 70 m. long and 17–23 m. wide island is one of the Fox group of the Aleutian Islands, SW Alaska.

Urup Island: The fourth largest of the Kurile Islands, Russian SFSR, it is 581 sq. m. in area.

Usedom: This 172 sq. m. island in the Baltic is divided between Mecklenburg (East Germany) and Szczecin (Poland).

351

Vansittart Island: A 47 m. long and 6–16 m. wide island in Foxe Channel just off S Melville Peninsula, Northwest Territories.

Vanua Levu: Formerly called Sandalwood Island, this high island, the second largest of the Fiji Islands in the SW Pacific, is 2137 sq. m. in area. It is volcanic in origin.

Vesteralen: Also spelled Vesteraalen; the largest island in this Norwegian archipelago N of the Lofoten Islands is Hinnoy. The group has a total area of 1200 sq. m.

Visayan Islands: The main islands in this Philippines group are Bohol, Cebu, Leyte, Masbate, Negros, Panay, and Samar. Its location is between Luzon and Mindanao.

Wake Island: Although it fell to the Japanese in 1941, this 45 m. long and 2¼ m. wide island was recaptured by the U.S. in 1945 after heavy bombing. It is comprised of an atoll and three islets between Hawaii and Guam.

Watubela Islands (Watoebela): The largest islands of this group are Kasiui (9 m. long and 2 m. wide) and Tior (6 m. long and 3 m. wide). The group is part of the S Moluccas, Indonesia, between the Banda and Arafura Seas.

Wellesley Islands: The largest of this uninhabited group of islands off the NW coast of Queensland, Australia, in the Gulf of Carpentaria, is Mornington.

ISLANDS

Wellington Island: This island of S Chile is 100 m. long and 15–25 m. wide. It is separated from the Madre de Dios Archipelago by the Trinidad Gulf.

Wessel Islands: The largest of this 70 m. chain of islands is 30 m. long and 7 m. wide. Located in the Arafura Sea, they extend NE from Napier Peninsula, Northern Territory, Australia.

West Spitsbergen: This 15,000 sq. m. island is the largest of the Spitsbergen group in the Arctic Ocean.

Whidbey Island: A 40 m. long island in Puget Sound, NW Washington.

Woodlark Island: This volcanic island is 40 m. long and 10 m. wide. Its location is in the Territory of Papua, SW Pacific, 175 m. E of New Guinea.

Wrangell Island: This island lies between Etolin Island and the mainland of SE Alaska. Part of the Alexander Archipelago, it is 30 m. long and 5–14 m. wide.

Yaeyama-gunto: Sometimes referred to as Yagamaretto, this 247 sq. m. island group is the southernmost of the Ryukyu Islands.

Yaku-shima: This 208 sq. m. island belonging to Japan is 40 m. S of Cape Sata, Kyushu, in the East China Sea.

Zealand: This is the largest island of Denmark. It is 2709 sq. m. in area and its location is between the Kattegat and the Baltic Sea.

Acknowledgments

Islands have fascinated us since childhood. We have collected them, studied them, and alas, seen many of them "developed" and polluted within a matter of years. No two people could know all the islands of the world—not even such "islamaniacs" as we—so we owe a deep debt to the many scientists, historians, photographers, storytellers and travelers, painters and writers who have shared with us information and their infatuation with islands. We are also indebted to the many museums, librarians, and historical associations who have supplied information and pictures.

That great island lover, John Millington Synge, used to say that all art was a collaboration between a writer and the peoples of a place. We feel, too, that all art must be a collaboration between a writer and his friends—particularly those who share special skills and offer a very special response and patience to a project of several years' duration; as such, Susan Belcher is a very special collaborator indeed. To our sailing friends Benjamin Belcher, Harry and Carolyn Royal, we owe the gift of small, sweet islands caught at dawn or in twilight from their boats. Many persons have contributed to the preparation of this manuscript and we are particularly grateful on this score to Betty Shalders, Heather Mason, who compiled the glossary of islands, Margarita and Kate Siafaca, Despina Paisidou, and Jennifer Middleton.

Gogo Lewis and Ralph Paterline have given us the benefit of their wide pictorial experience, and John Marion has offered wide editorial experience. James Linton has also been invaluable as the manuscript went to press. Warren Infield's design has certainly contributed to our pleasure in the printed book.

The National Park Service of the Department of the Interior and the Fish and Wildlife Service together with the United States Geological Survey have been especially helpful. We are particularly grateful to Raine E. Bennett, Executive Director of the Islands Research Foundation. Dr. Henry Field, Anne Takashige, Mr. and Mrs. William Bedell, Gloria Heath, and Mr. and Mrs. Charles Wilds have volunteered exceptionally useful information. Once again, we would like to express our thanks to Dr. Robert Cushman Murphy, Lamont Curator Emeritus of Birds for the American Museum of Natural History, and Peter Kotsogean, the economic geographer, whose knowledge of islands is worldwide. Our lengthy stay on that green isle, Ireland, was enhanced by the kindness of Mr. and Mrs. Louis Mc-Sherry, Katharine Moran, and John Dillon.

The Underwater Society of America, particularly for the chapters on St. Thomas and St. Croix, have been knowledgeable and infinitely hospitable. We should particularly like to thank the Burnses and the Carsons of St. Croix. The late Cathy Carson contributed an enthusiasm as magical as one's first dive underwater.

Many new and antiquarian book shops supplied us with invaluable data as they have with many of our books. Mr. and Mrs. Philip Hubert, Jr., of the Sou' Wester Bookshop, Bellport, N.Y.; Mr. and Mrs. Gilbert Ball of the Downstreet Book Shop in Cold Spring Harbor, N.Y.; the Barrett Book Shop in Stamford, Conn.; William MacIntosh of the MacIntosh Book Shop in Sanibel, Fla.; Mark Thompson of Thompson's Book Store in Sea Cliff, N.Y.; (Mrs.) Teddy Bookman and Evelyn Erb of The Country Bumpkin in Locust Valley, N.Y.; the Parnassus Book Service in Cape Cod, Mass.; Mr. and Mrs. Robert C. Hunt of Hunt's Book Store, Huntington, N.Y.; Nancy Mullen, Corner Book Shop, Setauket, N.Y., and particularly Donald Cameron of the Greenwich Book Store, Greenwich, Conn.

Many libraries, too, have aided us beyond the call of duty: the Boston Public Library; the Bellport Library of Bellport, N.Y.; the Patchogue Library of Patchogue, N.Y.; the Ferguson Library in Stamford, Conn.; the Huntington Library, Huntington, N.Y.; the St. Croix Library of the Virgin Islands; the New York Public Library and particularly the late Gerald McDonnell, chief of the American History Room; and the libraries of Martha's Vineyard and New Providence.

We benefited once again from the kindness and wisdom of the Audubon Society; many chapters of the Nature Conservancy throughout the world; state conservation departments throughout the United States; the Atomic Energy Commission; the Department of the Army, Corps of Engineers; the United States Department of Commerce, Coast and Geodetic Services; the League of Women Voters; the American Society of Limnology and Oceanography; the Marine Technology Society; the United States Air Force, and the United States Coast Guard.

The credits for the many countries and organizations that have contributed pictures are found within the book. We thank them all. In addition, we thank the following persons and organizations: Dr. and Mrs. Richard Amelar; The American Museum of Natural History; the American

354

Philosophical Society; Mr. and Mrs. Carlos Auriema; Harry E. Aumack, Jr.; the Australian National Travel Association; the Australian News and Information Bureau; F. Barnes, Regional Historian, National Park Service, Philadelphia; Russell Benedict; the Bermuda News Bureau; the British Information Bureau; George Butler; J. Robert Bromley; Preston Brown; Mr. and Mrs. John Chadwick; Mr. and Mrs. Robert A. Carter; the City of Refuge National Historical Park, Kona, Hawaii; Dan Dailey; Mr. and Mrs. Joseph Dietrich; Ernest Dodge of the Peabody Museum, Salem, Mass.; Laura G. Ebell; Encyclopedia Britannica Films, Inc.; Susan Fujitani; Nicholas Fisher; the Geological Survey of Canada, Ottawa; Mr. and Mrs. James Givens; Globe-Union; Hawaii Visitors Bureau; Hawaii Volcanoes National Park; Robert Higdon; Maria-Eugenia Huneeus; Mr. and Mrs. Charles Hvolbeck; the Irish Consulate; Isle Royale National Park; Mrs. James Jump; Mr. and Mrs. Roy Johnson; Mr. and Mrs. Don Jones; Lew King; The Library of Congress; the Long Island Historical Society, Brooklyn; Louisiana State University, Coastal Studies Institute; Martha MacGregor; George B. Magin, Jr.; the Maine Department of Economic Development; Mr. and Mrs. Frank B. Manley; Mariners Museum, Newport News, Va.; Jane Miller of Abraham & Strauss; the Museum of Fine Arts, Boston, Mass.; the National Aeronautics and Space Administration; the Geographical Survey of Canada; the National Archives Files; the National Film Board of Canada; The National Gallery of Canada, Ottawa; the National Library in Australia; the National Maritime Museum, England; the New York Academy of Sciences; the New York Folklore Society; the New York Historical Society; the New York State Historical Association, Cooperstown, N.Y.; the New York State Museum and Science Service; the *New York Times Magazine*; Dr. and Mrs. Harry Palevsky; Mr. and Mrs. George Pederson-Krag; the Provincial Archives, Victoria, Canada; Mr. and Mrs. Richard Ricardo; the San Juan National Historic Site, San Juan, P.R.; the San Juan, ·Puerto Rico, Office of Economic Development; the Scripps Institute of Oceanography; Philip Shay; Mr. and Mrs. Robert Sheffield; the Smithsonian Institution, Washington, D.C.; Joseph Spector; the Standard Oil Company of New Jersey; the State Historical Society of Wisconsin; Steuben Glass; the Superintendent of Documents, Washington, D.C.; John Swift; William Terry· the Union Title Insurance and Trust Company, San Diego, Calif.; U.S. Army Engineers; U.S. Coast Guard; U.S. Department of the Interior; U.S. Geological Survey; U.S. National Park Service; U.S. Post Office, West Palm Beach, Fla.; Universitetets Oldsaksamling, Oslo, Norway; University of California at San Diego; the University Museum of Philadelphia; the University Museum of Zoology, Cambridge, England; Ted Valpey; Virgin Islands National Park, St. Thomas; Peter Wainwright; Joan Walsh; Mr. and Mrs. George Webster; Western Electric; H. L. White; Rex Whitnack; George Wilcox; Dr. and Mrs. Clark Williams; Caroline P. Wister; Woods Hole Marine Biological Laboratory; and Martin Zwerin.

Selected Bibliography

Abbot, Charles G. (ed.), *The Smithsonian Series*. New York: The Series Publishers, Inc., 1931.

Abbott, R. Tucker, *Sea Shells of the World: A Guide to the Better-Known Species*. New York: Golden Press, 1962.

A Guide to Key West. (Sponsored by the Florida State Planning Board.) New York: Hastings House, Publishers, 1941.

Amory, Cleveland, *The Last Resorts*. New York: Grosset & Dunlap, 1952.

Amory, Cleveland, *Who Killled Society?* New York: Pocket Books, Inc., 1960.

Amos, William H., *The Life of the Seashore*. New York: McGraw-Hill Book Company, 1966.

Amundsen, Roald, *My Life as an Explorer*. New York: Doubleday, Page, 1927.

Anderson, Bern, *The Life and Voyages of Captain George Vancouver. Surveyor of the Sea*. Washington: University of Washington Press, 1960.

Anderson, Charles R., *Melville in the South Seas*. New York: Dover Publications, Inc., 1939.

A Proposed National Seashore Field Investigation Report. United States Department of the Interior—National Park Service, 1958.

Arnold, Augusta F., *The Sea-Beach at Ebb-Tide*. New York: Dover Publications, Inc., 1968.

Ashton, John, *The Dawn of the XIXth Century in England*. New York: G. P. Putnam's Sons, N.D.

Avril, Gilles, *The Conquest of Sea*. London: Burke Publishing Co. Ltd., 1960.

A World of Wonders or *Marvels in Animate and Inanimate Nature*, New York: Appleton and Company, 1881.

Baarslag, Karl, *Coast Guard to the Rescue*. New York: Farrar & Rinehart, Inc., 1937.

Babcock, William H., *Legendary Islands of the Atlantic*. New York: American Geographical Society, 1922.

Backhouse, James, *A Narrative of a Visit to the Australian Colonies*. London: Hamilton, Adams, and Co., 1843.

Ballantyne, M. R., *The Coral Island*. London: Andrew Dakers Limited, N.D.

Barber, John W., *Historical Collections of New Jersey: Past and Present*. New Haven, Conn.: Published by John W. Barker for Justus H. Bradley, 1868.

Barrett, Richmond, *Good Old Summer Days*. Boston: Houghton Mifflin Company, 1952.

Bascom, Willard, *Waves and Beaches: The Dynamics of the Ocean Surface*. Garden City, N.Y.: Doubleday & Company, Inc., 1964.

Bates, Marston, *The Nature of Natural History*. New York: Charles Scribner's Sons, 1950.

Bates, Marston, *Where Winter Never Comes*. New York: Charles Scribner's Sons, 1952.

Batton, Louis J., *The Nature of Violent Storms*. Garden City, N.Y.: Doubleday & Company, Inc., 1961.

Beck, Horace P., *The Folklore of Maine*. Philadelphia: J. B. Lippincott Company, 1957.

Beebe, William, *Exploring With Beebe*. New York, London: G. P. Putnam's Sons, 1932.

Beebe, William, *Nonsuch: Land of Water*. New York: Blue Ribbon Books, Inc., 1932.

Beebe, William, *The Log of the Sun: A Chronicle of Nature's Year*. Garden City, N.Y.: Garden City Publishing Company, Inc., 1927.

Beer, Sir Gavin de, *Charles Darwin, a Scientific Biography*. New York: Doubleday & Company, Inc., 1965.

Bell, Euphemia Y. and Associates, *Beautiful Bermuda*. New York & Bermuda: Beautiful Bermuda Publishing Co., Inc., 1947.

Bent, Arthur Cleveland, *Life Histories of North American Gulls and Terns*. New York: Dover Publications, Inc., 1963.

Bent, Arthur Cleveland, *Life Histories of North American Shore Birds, Part I*. New York: Dover Publications, Inc., 1962.

Berrill, N. J., *The Living Tide*. New York: Fawcett Publications, Inc., 1956.

Berrill, N. J., and Berrill, Jacquelyn, *1001 Questions Answered About the Seashore*. New York: Grosset & Dunlap, 1957.

Beston, Henry, *The Outermost House*. New York: The Viking Press, 1962.

357

ISLANDS

Bibby, Geoffrey, *The Testimony of the Spade*. New York: Alfred A. Knopf, 1956.

Birket-Smith, Kaj, *Primitive Man and His Ways*. Cleveland & New York: The World Publishing Company, 1957.

Blacker, Irwin R. (ed.), *Hakluyt's Voyages*. New York: The Viking Press, 1965.

Blake, Frances, *The Dolphin Guide to Cape Cod, Martha's Vineyard and Nantucket*. New York: Doubleday & Company, Inc., 1964.

Boles, Paul D., *Parton's Island*. New York: The Macmillan Company, 1958.

Boulter, B. C., *The Pilgrim Shrines of England*. London: Philip Allan & Co., Ltd., 1928.

Bradford, Ernle, *The Greek Islands*. New Yorrk: Harper & Row, Publishers, Incorporated, 1966.

Bradford, Gamaliel, *Darwin*. New York: Houghton Mifflin Company, 1926.

Brenan, Gerald, *South From Granada*. New York: Grove Press, Inc., N.D.

Briggs, Peter, *Water, the Vital Essence*. New York: Harper & Row, Publishers, 1967.

Brown, Henry C., *The Story of Old New York*. New York: E. P. Dutton & Co., Inc., 1934.

Brown, Henry C. (ed.), *Valentine's Manual of Old New York*. New York: Valentine's Manual, Inc., 1926.

Bulfin, William, *Rambles in Eirinn*. Dublin: M. H. Gill & Son, Ltd., 1929.

Bullen, Frank T., *Our Heritage the Sea*. New York: E. P. Dutton and Company, 1907.

Burgess, Robert H., *This Was Chesapeake Bay*. Cambridge, Maryland: Cornell Maritime Press, Inc., 1963.

Callender, James H., *Yesterdays in Little Old New York*. New York: Dorland Press, 1929.

Cameron, Ian, *Lodestone and Evening Star: The Epic Voyages of Discovery 1493 B.C.–1896 A.D.* New York: E. P. Dutton & Co., Inc., 1966.

Campbell, Alexander, *The Heart of Japan*. New York: Alfred A. Knopf, 1961.

Carloquist, Sherwin, *Island Life*. New York: The Natural History Press, 1965.

Carlova, John, *Mistress of the Seas*. New York: The Citadel Press, 1964.

Carpenter, E. S. (ed.), *Explorations Study in Culture and Communication*. Toronto, Canada: University of Toronto, 1953.

Carr, Archie, *The Windward Road*. New York: Alfred A. Knopf, 1956.

Carson, Rachel L., *The Sea Around Us*. New York: Oxford University Press, 1951.

Cary, Richard, *Sarah Orne Jewett*. New Haven: College and University Press, 1962.

Chadwick, John, *The Decipherment of Linear B*. New York: Random House, 1958.

Chamberlain, Barbara B., *These Fragile Outposts*. New York: The Natural History Press, 1964.

Chapin, Henry, and Smith, F. G. Walton, *The Ocean River*. New York: Charles Scribner's Sons, 1962.

Chapman, Esther, *Pleasure Island: The Book of Jamaica*. Kingston, Jamaica: The Arawak Press, 1955.

Chase, Owen, *Shipwreck of the Whaleship Essex*. New York: Corinth Books, 1963.

Clarke, Arthur, and Wilson, Mike, *The Treasure of the Great Reef*. New York: Harper & Row, Publishers, 1964.

Coker, R. E., *This Great and Wide Sea: An Introduction to Oceanography and Marine Biology*. New York: Harper & Brothers, 1962.

Colvin, Sidney (ed.), *The Letters of Robert Louis Stevenson to His Family and Friends*. New York: Charles Scribner's Sons, 1899.

Connett, Eugene V. (ed.), *Duck Shooting Along the Atlantic Tidewater*. New York: William Morrow and Co., Inc., 1947.

Conover, David, *Once Upon an Island*. New York: Crown Publisher, 1967.

Cooper, Elizabeth K., *Science on the Shores and Banks*. New York: Harcourt, Brace & World, Inc., 1960.

Cottrell, Leonard, *Realms of Gold*. Greenwich, Conn.: New York Graphic Society Publishers, Ltd., 1963.

Coulton, G. G., *Medieval Faith and Symbolism. Part I of Art and the Reformation*. New York: Harper & Brothers, Publishers, 1958.

Cousteau, Jacques-Yves, and Dugan, James (eds.), *Captain Cousteau's Underwater Treasury*. New York, Evanston, London: Harper & Row, Publishers, 1959.

Covarrubias, Miguel, *Island of Bali*. New York: Alfred A. Knopf, 1956.

Cowen, Robert C., *Frontiers of the Sea*. New York: Bantam Books, 1963.

Crone, G. R. (ed.), *The Explorers*. New York: Thomas Y. Crowell Company, 1962.

Cunningham, John T., *The New Jersey Shore*. New Brunswick, N.J.: Rutgers University Press, 1958.

Dale, Paul W. (ed.), *Seventy North to Fifty South*. New Jersey: Prentice-Hall, Inc., 1969.

Dampier, William, *A New Voyage Round the World*. New York: Dover Publications, Inc., 1968.

Dana, James D., *The Geological Story Briefly Told*. New York: Ivison, Blakeman, Taylor and Company, 1875.

Danielsson, Bengt, *The Happy Island*. London: George Allen & Unwin Ltd., 1954.

Darling, F. Fraser, *Island Years*. London: G. Bell and Sons, Ltd., 1944.

Darling, F. Fraser, and Milton, John P. (eds.), *Future Environments of North America*. Garden City, N.Y.: The Natural History Press, 1966.

359

ISLANDS

Darling, Louis, *The Gull's Way*. New York: William Morrow and Company, 1965.

Darwin, Charles, *Coral Reefs*. Los Angeles: University of California Press, 1962.

Darwin, Charles, *The Voyage of the Beagle*. New York: Bantam Books, 1958.

Day, A. Grove, *Hawaii and Its People*. New York: Duell, Sloan and Pearce, 1955.

Day, Bunny, *Catch 'Em and Cook 'Em*. New York: Doubleday & Company, Inc., 1961.

Deacon, G. E. R. (ed.), *Seas, Maps, and Men*. London: Geographical Projects Ltd., 1962.

Deane, Shirley, *In a Corsican Village*. New York: The Vanguard Press, Inc., 1965.

Defant, Albert, *Ebb and Flow: The Tides of Earth, Air, and Water*. Ann Arbor, Michigan: The University of Michigan Press, 1958.

Delaney, Edmund T., *Greenwich Village*. Massachusetts: Barre Publishers, 1968.

De la Rue, Aubert E., *Man and the Winds*. New York: Philosophical Library, 1955.

Dixon, Peter L., *Men and Waves*. New York: Ballantine Books, 1966.

Dodd, Dorothy, *Florida the Land of Romance*. Tallahassee: The Peninsular Publishing Co., 1957.

Dodge, Ernest S., *New England and the South Seas*. Cambridge, Massachusetts: Harvard University Press, 1965.

Dodge, Ernest S., *Northwest by Sea*. New York: Oxford University Press, 1961.

Donnelly, Ignatius, *Atlantis: The Antediluvian World*. New York: Gramercy Publishing Co., 1949.

Douglas, Marjory S., *Florida: The Long Frontier*. New York: Harper & Row, Publishers, 1967.

Douglas, Norman, *South Wind*. New York: Grosset & Dunlap, 1952.

Downey, Joseph T., *The Cruise of the Portsmouth, 1845–1847: A Sailor's View of the Naval Conquest of California*. New Haven, Connecticut: Yale University Press, 1963.

Dumas, Alexandre, *The Count of Monte-Cristo*. New York: Grosset & Dunlap Publishers, 1946.

Dunbar, Carl O., *Historical Geology*. New York: John Wiley & Sons, Inc., 1949.

Eckert, Allan W., *The Great Auk*. Boston: Little, Brown and Company, 1963.

Edwards and Williams (eds.), *The Great Famine—Studies in Irish History 1945–52*. New York: New York University Press, 1957.

Ellison, Joseph W., *Tusitala of the South Pacific*. New York: Hastings House, Publishers, 1953.

Emerson, Edwin, Jr., A *History of the Nineteenth Century Year by Year*. New York: P. F. Collier and Son, 1900.

Engel, Leonard (and the Editors of *Life*), *The Sea*. New York: Time Inc., 1961.

Espy, James P., *The Philosophy of Storms*. Boston: Charles C. Little and James Brown, 1841.

Evans, I. O., *The Observer's Book of Sea and Seashore*. London: Frederick Warne & Co. Ltd., 1964.

Eydoux, Henri P., *History of Archaeological Discoveries*. London: Leisure Arts Limited Publishers, 1966.

Fairchild, David, *Garden Islands of the Great East*. New York: Charles Scribner's Sons, 1944.

Farb, Peter, *Face of North America: The Natural History of a Continent*. New York: Harper & Row, Publishers, Incorporated, 1963.

Farb, Peter (and the Editors of *Life*), *Ecology*. New York: Time Incorporated, 1963.

Farnham, E. C. Joseph (ed.), *Brief Historical Data and Memories of My Boyhood Days in Nantucket*. Rhode Island: Joseph E. C. Farnham, 1923.

Fellows, Henry P., *Boating Trips on New England Rivers*. Boston: Cupples, Upham, and Company, 1884.

Fermor, Patrick L., *The Traveller's Tree*. New York: Harper & Brothers Publishers, N.D.

Field, Henry, *The Track of Man*. London: Peter Davies, 1955.

Figuier, Louis (revised by Wright, E. P.), *The Ocean World*. New York: D. Appleton and Co., N.D.

Fisher, James; Simon, Noel, and Vincent, Jack, *Wildlife in Danger*. New York: The Viking Press, 1969.

Fiske, John, *The Beginnings of New England*. Boston & New York: Houghton Mifflin Company, 1889.

Freuchen, Peter, *Book of the Eskimos*. Cleveland and New York: The World Publishing Company, 1961.

Fritz, Florence, *Unknown Florida*. Coral Gables, Florida: University of Miami Press, 1963.

Froude, James A., *The English in the West Indies*. New York: Charles Scribner's Sons, 1900.

Gaddis, Vincent, *Invisible Horizons: True Mysteries of the Sea*. Philadelphia: Chilton Company, 1965.

Garse, Robert, *Keepers of the Lights*. New York: Charles Scribner's Sons, 1969.

Gaskell, T. F., *World Beneath the Oceans*. Garden City, N.Y.: The Natural History Press, 1964.

Gauguin, Paul, *Noa Noa*. New York: Nicholas L. Brown, 1920.

Gaul, Albro, *The Wonderful World of the Seashore*. New York: Appleton-Century-Crofts, Inc., 1955.

361

ISLANDS

Gehring, Mabel G., *On a Scottish Island*. Cleveland and New York: The World Publishing Company, 1949.

Geismar, Maxwell (ed.), *The Whitman Reader*. New York: Pocket Books, Inc., 1955.

Gibbins, Robert, *John Graham Convict 1824*. New York: A. S. Barnes & Company, 1957.

Gibbons, Euell, *Beachcomber's Handbook*. New York: David McKay Company, Inc., 1967.

Giddings, Louis J., *Ancient Men of the Arctic*. New York: Alfred A. Knopf, 1967.

Gilbreth, Frank B., Jr., *Inside Nantucket*. New York: Thomas Y. Crowell Company, 1954.

Gilchrist, David T. (ed.), *The Growth of the Seaport Cities 1790–1825*. Virginia: Eleutherian Mills-Hagley Foundation, 1967.

Gleason, Duncan, *The Island and Ports of California*. New York: The Devin-Adair Company, 1958.

Goodman, Henry (ed.), *The Selected Writings of Lafcadio Hearn*. New York: The Citadel Press, 1949.

Goodman, Nathan G. (ed.), *A Benjamin Franklin Reader*. New York: Thomas Y. Crowell Company, 1945.

Goodrich, Lloyd, *Winslow Homer*. New York: George Braziller, Inc., 1959.

Gordon, John (ed.), Hills, L. Rust (ed.), *New York New York*. New York: Shorecrest, Inc., 1965.

Gosse, Philip H., *The History of Piracy*. New York: Tudor Publishing Company, 1932.

Gosse, Philip H., *Year at the Shore*. London: Alexander Strahan Publisher, 1865.

Gottmann, Jean, *Megalopolis*. Cambridge, Massachusetts: The M.I.T. Press, 1964.

Graves, Charles (ed.), *Fourteen Islands in the Sun*. New York: Hart Publishing Co., 1965.

Greely, Adolphus W., *Three Years of Arctic Service*. New York: Charles Scribner's Sons, 1886.

Greene, David H., and Stephens, M. Edward, *J. M. Synge—1871–1909*. New York: The Macmillan Company, 1959.

Gwynn, Stephen, *Highways and Byways in Donegal and Antrim*. London: MacMillan and Co., 1899.

Hall, Henry Marion, *A Gathering of Shore Birds*. New York: Bramhall House, 1960.

Hall, James N., *The Forgotten One and Other True Tales of the South Seas*. Boston: Little, Brown and Company, 1950.

Hall, James N., and Nordhoff, Charles B., *Faery Lands of the South Seas*. Garden City, New York: Garden City Publishing Co., Inc., 1926.

Halsey, Francis W., *Seeing Europe With Famous Authors*. New York and London: Funk & Wagnalls Company, 1914.

Hamlyn, Paul, *Greek Mythology*. London: Paul Hamlyn Limited, 1963.

362

Hapgood, Charles H., *Maps of the Ancient Sea Kings*. Philadelphia and New York: Chilton Book Company, 1966.

Harrisson, Tom, *Savage Civilization*. New York: Alfred A. Knopf, 1937.

Harvey, D. C. (ed.), *Journeys to the Island of St. John or Prince Edward·Island, 1775–1832*. Toronto: The Macmillan Company of Canada Limited, 1955.

Hatcher, Harlan, and Walter, Frich A., *A Pictorial History of the Great Lakes*. New York: Bonanza Books, 1969.

Haverty, Martin, *The History of Ireland*. New York: Thomas Kelly, Catholic Publisher, 1867.

Havighurst, Walter (ed.), *Land of the Long Horizons*. New York: Coward-McCann Inc., 1960.

Hawkes, Jacquetta, and Leonard, Sir Woolley, *History of Mankind*, Vol. I.: *Prehistory and the Beginnings of Civilization*. New York: Harper & Row, Publishers, N.D.

Hawthorne, Daniel, and Minot, Francis, *The Inexhaustible Sea*. New York: Collier Books, 1961.

Hawthorne, Julian, *Nathaniel Hawthorne and His Wife*. Vol. II. Boston and New York: Houghton Mifflin and Company, 1896.

Hay, John, *The Run*. Garden City, N.Y.: Doubleday & Company, Inc., 1965.

Hay, John, and Farb, Peter, *The Atlantic Shore: Human and Natural History From Long Island to Labrador*. New York: Harper & Row, Publishers, Incorporated, 1966.

Hays, H. R., *From Ape to Angel*. New York: Alfred A. Knopf, 1958.

Hayward, John (ed.), *Swift. Gulliver's Travels and Selected Writings in Prose & Verse*. New York: Random House, 1934.

Hazard, Patrick D., *The Dolphin Guide to Hawaii*. Garden City, N.Y.: Doubleday & Co., Inc., 1965.

Healy, J. Rev., *Irish Essays*. Dublin: Catholic Truth Society of Ireland, 1908.

Hemstreet, Charles, *Nooks and Corners of Old New York*. New York: Charles Scribner's Sons, 1809.

Hemstreet, Charles, *The Story of Manhattan*. New York: Charles Scribner's Sons, 1901.

Henle, Fritz, *Virgin Islands*. New York: Hastings House, 1949.

Hennessy, Pope James, *West Indian Summer*. London: B. T. Batsford Ltd., 1943.

Herrman, Paul, *The Great Age of Discovery*. New York: Harper & Brothers Publishers, 1958.

Heston, A. M., *Absegami: Annals of Eyren Haven and Atlantic City, 1609 to 1904*. Camden, N.J.: Sinnickson Chew & Sons Co., 1904.

Holder, Charles F., *The Channel Islands of California*. Chicago: A. C. McClurg & Co., 1910.

Holmes, Oliver W., *The Autocrat of the Breakfast-Table*. Boston and New York: Houghton Mifflin and Company, 1892.

Hough, Beetle H. (ed.), *Vineyard Gazette Reader*. New York: Harcourt, Brace & World, Inc., 1967.

363

ISLANDS

Hough, Edith L., *Sicily*. New Hampshire: Marshall Jones Company, N.D.
Hudson, W. H., *Afoot in England*. London: J. M. Dent and Sons Ltd., 1909.
Hulme, F. E., *Natural History: Lore and Legend*. London: Bernard Quaritch, 1895.
Hutchinson, Thomas J., *Two Years in Peru*. London: Sampson Low, Marston, Low, & Searle, 1873.
Huxley, Aldous, *Island*. New York: Bantam Books, Inc., 1962.
Huxley, Anthony (ed.), *Oceans and Islands*. New York: G. P. Putnam's Sons, 1962.
Hyde, Walter W., *Ancient Greek Mariners*. New York: Oxford University Press, 1947.
Hyman, Stanley E., *The Tangled Bank. Darwin, Marx, Frazer and Freud as Imaginative Writers*. New York: Atheneum, 1962.

Ingram, H. J., *The Islands of England*. London: B. T. Batsford, Ltd., 1952.
Introducing the British Pacific Islands. London: Her Majesty's Stationery Office, 1951.
Islands in the Sun. New York: Islands in the Sun Club, 1960.

Janvier, Thomas A., *In Old New York*. London: Harper & Brothers Publishers, 1894.
Jewett, Sarah O., *The Country of the Pointed Firs and Other Stories*. New York: Doubleday & Co., Inc., 1956.
John, Bruce St. (ed.), *John Sloan's New York Scene*. New York: Harper & Row, Publishers, Inc., 1965.
Johnson, Clifton, *New England and Its Neighbors*. New York: The Macmillan Company, 1902.
Johnson, E. D. H., *The Poetry of Earth*. New York: Atheneum, 1966.
Johnson, E. Pauline, *Legends of Vancouver*. Toronto: McClelland & Stewart Limited, 1961.
Johnstone, Kathleen Yerger, *Sea Treasure*. Boston: Houghton Mifflin Company, 1957.
Jones, Parry D., *Welsh Legends and Fairy Lore*. London: B. T. Batsford Ltd., 1953.
Joyce, W. P., *Atlas and Cyclopedia of Ireland. Part I*. Sullivan, M. A., and Nunan, P. P., *The General History. Part II*. New York: Murphy & MacCarthy, Publishers, 1900.

Kahn, E. J., Jr., *A Reporter in Micronesia*. New York: W. W. Norton & Company, Inc., 1966.
Kane, Robert S., *South Pacific A to Z*. Garden City, N.Y.: Doubleday & Company, Inc., 1966.
Kardiner, Abram, and Preble, Edward, *They Studied Man*. New York: The World Publishing Co., 1961.
Keast, Allen, *Australia and the Pacific Islands*. New York: Random House, 1966.

Keesing, Felix M., *The South Seas in the Modern World.* New York: The John Day Company, 1946.

Kellogg, Winthrop N., *Porpoises and Sonar.* Chicago: Phoenix Books, The University of Chicago Press, 1965.

Kemp, P. K., and Lloyd, Christopher, *Brethren of the Coast: Buccaneers of the South Seas.* New York: St. Martin's Press, 1960.

Kendrick, T. D., *The Druids: A Study in Keltic Prehistory.* New York: R. V. Coleman, N.D.

Kennedy, Stetson, *Palmetto Country.* New York: Duell, Sloan & Pearce, 1942.

Kieran, John, *A Natural History of New York City.* Boston: Houghton Mifflin Company, 1959.

Kimbrough, Emily, *Water, Water Everywhere.* New York: Harper & Row, Publishers, 1956.

King, Cuchlaine A. M., *Beaches and Coasts.* London: Edward Arnold Ltd., 1961.

Kingsley, Charles, *At Last: A Christmas in the West Indies.* London: Macmillan and Co., 1873.

Kipling, Rudyard, *Land and Sea Tales for Scouts and Scout Masters.* New York: Doubleday, Page & Company, 1924.

Klein, David, and King, Mary Louise, *Great Adventures in Small Boats.* New York: Collier Books, 1963.

Knight, Richard P., *The Symbolical Language of Ancient Art and Mythology.* New York: J. W. Bouton, 706 Broadway, 1876.

Krutch, Joseph W., *The World of Animals.* New York: Simon and Schuster, 1961.

Kubly, Herbert, *Easter in Sicily.* New York: Simon and Schuster, 1956.

Kuenen, P. H., *Realms of Water: Some Aspects of its Cycle in Nature.* New York: John Wiley & Sons, Inc., 1963.

Kurath, Hans, *A Word Geography of the Eastern United States.* U.S.A.: The University of Michigan Press, 1949.

La Nature, Revue des Sciences, 1884, Paris, France.

Larousse Encyclopedia of Mythology. New York: Prometheus Press, 1959.

Latil, Pierre de, and Rivoire, Jean, *Sunken Treasure.* New York: Hill and Wang, 1962.

Lawson, John C., *Modern Greek Folk Lore and Ancient Greek Religion.* Cambridge: University Press, 1910.

Laycock, George, *The Sign of the Flying Goose.* New York: The Natural History Press, 1965.

Lee, Storrs W. (ed.), *Maine.* New York: Funk & Wagnalls, 1965.

Lee, Willis T., *The Face of the Earth As Seen From the Air.* New York: American Geographical Society, 1922.

Leenhardt, Maurice, *The World's Art: Folk Art of Oceania.* Paris: Les Editions du Chene, 1950.

ISLANDS

Lehane, Brendan, *The Quest of Three Abbots*. London: For John Murray, by the Camelot Press Ltd., 1968.

Lehner, Ernest and Johanna, *How They Saw the New World*. New York: Tudor Publishing Company, 1966.

Leslie, Robert C., *A Waterbiography*. London: Chapman & Hall, Ltd., 1894.

Lethbridge, T. C., *Herdsmen & Hermits: Celtic Seafarers in the Northern Seas*. Cambridge: Bowes and Bowes Publishers Limited, 1950.

Lewisohn, Florence, *Tales of Tortola and the British Virgin Islands*. Alroy Printing Company, 1966.

Leyda, Jay, *The Melville Log*. New York: Harcourt, Brace and Company, 1951.

Lilly, John C., *Man and Dolphin*. New York: Pyramid Publications, Inc., 1962.

Lissner, Ivar, *Man, God and Magic*. New York: G. P. Putnam's Sons, 1961.

Lobeck, Armin K., *Things Maps Don't Tell Us*. New York: The Macmillan Company, 1958.

Lonsdale, Adrian L., and Kaplan, H. R., *A Guide to Sunken Ships in American Waters*. Arlington, Va.: Compass Publications, Inc., 1964.

Lovell, Caroline C., *The Golden Isles of Georgia*. Boston: Little, Brown and Company, 1932.

Lovett, Richard, *Ireland Illustrated With Pen and Pencil*. New York: Hurst & Company, 1891.

Lowe, Percy R., *A Naturalist on Desert Islands*. London: Witherby & Co., 1911.

Lunt, Dudley C., *The Woods and the Sea*. New York: Alfred A. Knopf, 1965.

Lurie, Richard, *Under the Great Barrier Reef*. London: Jarrolds Publishers, 1966.

MacCulloch, Cauon J. A. (ed.), *The Mythology of All Races. Vol. VIII*. Boston: Marshall Jones Co., 1929.

MacCurdy, Edward, *The Notebooks of Leonardo da Vinci*. New York: Reynal & Hitchcock, N.D.

Macdonald, Malcolm, *Canadian North*. London: Oxford University Press, 1945.

Mackendrick, Paul, *The Greek Stones Speak*. New York and Toronto: The New American Library, 1962.

Macleod, Kenneth, *The Road to the Isles*. Edinburgh: Robert Grant & Son, 1927.

Macmillan, Allister (ed.), *The West Indies Illustrated*. London: W. H. & L. Collingridge, 1909.

Mahn-Lot, Marianne, *Columbus*. New York: Grove Press, Inc., 1961.

Malinowski, Bronislaw, *A Diary in the Strict Sense of the Term*. New York: Harcourt, Brace & World, Inc., 1967.

Manley, Seon, *Long Island Discovery*. Garden City, N.Y.: Doubleday & Company, Inc., 1966.

Mann, William M., *Wild Animals In and Out of the Zoo*. New York: The Series Publishers, Inc., 1930.

Maraini, Fosco, *The Island of the Fisherwomen*. New York: Harcourt, Brace & World, Inc., 1960.

Marsden, Christopher, *The English at the Seaside*. London: Collins, 1947.

Martin, Sidney W., *Florida's Flagler*. Athens: The University of Georgia Press, 1949.

Matthews, William H., III, *Fossils: An Introduction of Prehistoric Life*. New York: Barnes & Noble, Inc., 1964.

Maury, Matthew F., *The Physical Geography of the Sea and its Meteorology*. Cambridge, Mass.: The Belknap Press of Harvard University Press, 1963.

Mavor, James W., Jr., *Voyage to Atlantis*. New York: G. P. Putnam's Sons, 1969.

Maxwell, Gavin, *People of the Reeds*. New York: Harper & Brothers Publishers, 1957.

Mayoux, Jean-Jacques, *Melville*. New York: Grove Press, Inc., 1960.

McAdam, Roger Williams, *Salts of the Sound*. New York: Stephen Daye Press, 1957.

McCabe, James D., *New York by Sunlight and Gaslight*. New York: Union Publishing House, 1885.

McCampbell, Coleman, *Texas Seaport: The Story of the Growth of Corpus Christi and the Coastal Bend Area*. New York: Exposition Press, 1952.

McCarthy, Justin (editor in chief), *Irish Literature*. New York: Bigelow, Smith & Company, 1904.

McCullough, Esther M. (ed.), *As I Pass, O Manhattan*. New York: Coley Taylor, Inc., N.D.

McDonnell, Randal, *When Cromwell Came to Drogheda*. Dublin: M. H. Gill & Son, Ltd., N.D.

McLaren, Moray, *The Highland Jaunt*. London: Jarrolds Publishers, Ltd., 1954.

Mead, Margaret, *Coming of Age in Samoa*. New York: Dell Publishing Co., Inc., 1928.

Mégroz, R. L., *Profile Art Through the Ages*. New York: Philosophical Library, Inc., 1949.

Melville, Herman, *Moby Dick*. Boston: Page & Company, 1892.

Melville, Herman, *Typee*. Boston: L. C. Page & Company, Publishers, 1892.

Menard, Wilmon, *The Two Worlds of Somerset Maugham*. California: Sherbourne Press, Inc., 1965.

Michener, James A., *Return to Paradise*. New York: Random House, 1950.

Millar, George, *Oyster River*. New York: Alfred A. Knopf, 1964.

Miller, Helen H., *Sicily and the Western Colonies of Greece*. New York: Charles Scribner's Sons, 1965.

Miller, Martin J., *The Martinique Horror and St. Vincent Calamity*. Philadelphia: Globe Bible Publishing Co., 1908.

Miller, Robert C., *The Sea*. New York: Random House, 1966.

Miner, Roy Waldo, *Field Book of Seashore Life*. New York: G. P. Putnam's Sons, 1950.

Mines, John F., A *Tour Around New York* and *My Summer Acre*. New York: Harper & Brothers, 1893.

Mitchell, Edwin V., *It's an Old State of Maine Custom*. New York: The Vanguard Press, Inc., 1949.

Mitchell, Joseph, *The Bottom of the Harbor*. Boston and Toronto: Little, Brown and Company, 1944.

Molony, Eileen (ed.), *Portraits of Islands*. London: Dennis Dobson Ltd., 1951.

Moore, Ruth, *Evolution*. New York: Time-Life Books, 1962.

Moorehead, Alan, *The Fatal Impact*. New York: Harper & Row, Publishers, 1966.

Morgan, Lloyd C., *Animal Sketches*. London: Edward Arnold, N.D.

Morison, Samuel E., *The Maritime History of Massachusetts 1783–1860*. Boston: Houghton Mifflin Company, 1961.

Morison, Samuel E., *The Story of Mount Desert Island Maine*. Boston: Little, Brown and Company, 1960.

Morison, Samuel E., and Obrecon, Mauricio, *The Caribbean as Columbus Saw It*. Boston: Little, Brown and Company, 1964.

Morton, J. E., *Molluscs*. New York: Harper & Brothers, 1958.

Munson, Gorham, *Penobscot: Down East Paradise*. Philadelphia: J. B. Lippincott Company, 1959.

Murphy, Robert, *A Certain Island*. New York: Avon Books, 1967.

Murphy, Robert C., *Fish-Shape Paumanok: Nature and Man on Long Island*. Philadelphia: The American Philosophical Society, 1964.

Murphy, Robert C., *Oceanic Birds of South America*. New York: The Macmillan Company, 1936.

Murray, Grace A., *Ancient Rites and Ceremonies*. London: Alston Rivers Ltd., 1929.

Murray, H. W., *The Hebrides*. New York: A. S. Barnes and Co., Inc., 1966.

Muspratt, Eric, *My South Sea Island*. England: Penguin Books Limited, 1931.

Nansen, Fridtjof, *The First Crossing of Greenland*. London: Longmans, Green, and Co., 1890.

Natural History. The Journal of the American Museum of Natural History. New York: The American Museum of Natural History, Volume XXVIII, 1928; Volume XXIX, 1929; Volume XXX, 1930; Volume XXXI, 1931; Volume XXXII, 1932.

Neale, Tom, *An Island to Myself*. New York: Avon Books, 1966.

Niering, William A., *The Life of the Marsh: The North American Wetlands.* New York: McGraw-Hill Book Company, 1966.

Norris, Kenneth S. (ed.), *Whales, Dolphins, and Porpoises.* Berkeley, Calif.: University of California Press, 1966.

O'Brien, Frederick, *Mystic Isles of the South Seas.* Garden City, N.Y.: Garden City Publishing Company, Inc., 1921.

O'Faolain, Sean, *The Silver Branch.* New York: The Viking Press, 1938.

Ogburn, Charlton, Jr., *The Winter Beach.* New York: William Morrow & Company, Inc., 1966.

Oliver, Douglas L., *The Pacific Islands.* New York: Doubleday & Company, Inc., 1961.

Ommanney, O. F., *Isle of Cloves: A View of Zanzibar.* Philadelphia and New York: J. B. Lippincott Company, 1956.

Otto, Walter F., *The Homeric Gods.* Boston: Beacon Press, 1954.

Ovington, Ray, *The Complete Guide to Fresh and Salt Water Fishing.* New York: Thomas Nelson & Sons, 1961.

Pacific Coast Recreation Area Survey. Washington, D.C.: U.S. Department of the Interior, National Park Service, 1959.

Parks for America. National Park Service, 1964.

Parks, George B., *Richard Hakluyt and the English Voyages.* New York: American Geographical Society, 1928.

Parry, J. H., and Sherlock, P. M., *A Short History of the West Indies.* New York: St. Martin's Press, 1966.

Payne, Robert, *The Island.* New York: Harcourt, Brace and Company, 1958.

Pearl, Richard M., *Geology.* (College Outline Series.) New York: Barnes & Noble, Inc., 1960.

Perry, W. J., *The Primordial Ocean.* London: Methuen & Co., Ltd., 1935.

Pettingill, Olin Sewall, Jr., *The Bird Watcher's America.* New York: McGraw-Hill Book Company, 1965.

Phillips, James D., *Salem and the Indies.* Boston: Houghton Mifflin Company, 1947.

Porter, Arthur K., *The Crosses and Culture of Ireland.* New Haven: Yale University Press, 1931.

Pratt, Theodore, *The Barefoot Mailman.* New York: Duell, Sloan and Pearce, 1961.

Preliminary Report: Laboratory Study of the Effect of an Uncontrolled Inlet on the Adjacent Beaches. Technical Memorandum No. 94 Beach Erosion Board Corps of Engineers, May, 1957.

Rasponi, Lanfranco, *The Golden Oases.* New York: G. P. Putnam's Sons, 1968.

Ray, Carleton, and Ciampi, Elgin, *The Underwater Guide to . . . Marine Life.* New York: A. S. Barnes and Company, Inc., 1956.

Reynolds, E. E., *Nansen.* Harmondsworth, Middlesex: Penguin Books, 1932.

369

Rich, Louise Dickinson, *The Coast of Maine*. New York: Thomas Y. Crowell Company, 1956.

Richardson, Wyman, *The House on Nauset Marsh*. New York: W. W. Norton & Company, Inc., 1955.

Rider, Fremont (ed.), *Rider's New York City*. New York: Henry Holt and Company, 1923.

Roberts, Adolphe W., *Jamaica, the Portrait of an Island*. New York: Coward-McCann, Inc., 1955.

Robertson, B. R., *Of Whales and Men*. New York: Alfred A. Knopf, 1954.

Rodman, Selden, *The Caribbean*. New York: Hawthorn Books, Inc., Publishers, 1968.

Rowlands, John J., *Spindrift: From a House by the Sea*. New York: W. W. Norton & Company, Inc., 1960.

Russell, Franklin, *The Secret Islands*. New York: W. W. Norton & Company, Inc., 1965.

Russell, F. S., and Yonge, C. M., *The Seas*. London: Frederick Warne & Co., Ltd., 1963.

Ryan, W. P., *The Pope's Green Island*. Boston, Mass.: Small, Maynard & Co., N.D.

Samuels, Edward A., *Our Northern and Eastern Birds*. New York: R. Worthington, 1883.

Sanders, Ruth M., *Seaside England*. London: B. T. Batsford Ltd., 1951.

Sanger, Marjory B., *Mangrove Island*. New York: The World Publishing Company, 1963.

Saunders, John Richard, *The World of Natural History*. New York: Sheridan House, 1952.

Scammon, Charles M., *The Marine Mammals*. New York: Dover Publications, Inc., 1968.

Schmitt, Waldo L., *Crustaceans*. Ann Arbor, Mich.: The University of Michigan Press, 1965.

Schroeder, Robert E., *Something Rich and Strange*. New York: Harper & Row, Publishers, Incorporated, 1965.

Scott, Genio C., *Fishing in American Waters*. New York: Orange Judd Company, 1875.

Scott, Peter, *The Eye of the Wind*. Boston: Houghton Mifflin Company, 1961.

Sears, Paul B., *Charles Darwin, the Naturalist as a Cultural Force*. New York: Charles Scribner's Sons, 1950.

Self, Margaret C., *Irish Adventure*. New York: A. S. Barnes and Company, 1954.

Seopold, Aldo, *A Sand County Almanac and Sketches Here and There*. New York: Oxford University Press, 1949.

Shaler, N. S., *Sea and Land*. New York: Charles Scribner's Sons, 1894.

Shannon, Howard J., *The Book of the Seashore*. New York: Doubleday,

Doran & Company, Inc., 1935.

Shapiro, Harry L., *The Heritage of the Bounty*. New York: Simon and Schuster, 1936.

Sharp, Andrew, *Ancient Voyagers in the Pacific*. Baltimore: Md.: Penguin Books, Inc., 1957.

Shelley, Henry C., *Literary By-paths in Old England*. Boston: Little, Brown and Company, 1909.

Shepard, Francis P., *Submarine Geology*. New York: Harper & Row, 1963.

Sheraton, Mimi, *City Portraits*. New York: Harper & Row Publishers, 1962.

Simpson, Blantyre E., *The Robert Louis Stevenson Originals*. New York: Charles Scribner's Sons, 1913.

Skelton, Robin, and Clark, David R. (eds.), *Irish Renaissance*. Republic of Ireland: The Dolmen Press, 1965.

Skene, William F., *Celtic Scotland*. Edinburgh: David Douglas, 1887.

Skinner, Charles M., *Myths & Legends of Our Own Land*. Vols. I and II. Philadelphia and London: J. B. Lippincott Company, 1896.

Smith, Bradley, *Escape to the West Indies*. New York: Alfred A. Knopf, 1956.

Smith, W. R., *Myths & Legends of the Australian Aboriginals*. London: George G. Harrap & Company, Ltd., 1930.

Smyth, Lindley (ed.), *By Earthquake and Fire*. Philadelphia: The International Press, N.D.

Snow, Edward R., *New England Sea Tragedies*. New York: Dodd, Mead & Company, 1960.

Snow, Edward R., *True Tales of Buried Treasure*. New York: Dodd, Mead & Company, 1963.

Spectorsky, A. C. (ed.), *The Book of the Earth*. New York: Appleton-Century-Crofts, Inc., 1957.

Spence, Lewis, *The History of Atlantis*. London: Rider & Co., N.D.

Squier, George E., *Peru: Incidents of Travel and Exploration in the Land of the Incas*. New York: Harper & Brothers, Publishers, 1877.

Starkey, Marion L., *Land Where Our Fathers Died*. Garden City, N.Y.: Doubleday & Company, Inc., 1962.

Stevenson, Elizabeth, *Lafcadio Hearn*. New York: The Macmillan Company, 1961.

Stevenson, Fanny and Robert Louis, *Our Samoan Adventure*. New York: Harper & Brothers, 1955.

Stevenson, Robert Louis, *Treasure Island*. U.S.A.: The New American Library of World Literature, 1965.

Stewart, Harris B., Jr., *The Global Sea*. Princeton, N.J.: D. Van Nostrand Company, Inc., 1963.

Stewart, John D., *Gibraltar the Keystone*. England: Louter Trend Co., 1967.

Stick, David, *The Outer Banks of North Carolina 1584–1958*. Chapel Hill, N.C.: The University of North Carolina Press, 1958.

Still, Bayrd, *Mirror for Gotham New York as Seen by Contemporaries from Dutch Days to the Present*. New York: University Press, 1956.

Stimson, J. Frank, *Songs and Tales of the Sea Kings.* Portland, Maine: The Anthoensen Press, 1957.

Stockdale, John, *Perouse's Voyage.* London: 1800.

Stockton, Frank R., *Buccaneers and Pirates of Our Coasts.* New York: Looking Glass Library, 1960.

Stone, Witmer, *Bird Studies at Old Cape May: An Ornithology of Coastal New Jersey,* Volumes I and II. New York: Dover Publications, Inc., 1965.

Stonehouse, Bernard, *Wideawake Island: The Story of the B.O.U. Centenary Expedition to Ascension.* London: Hutchinson & Co., Ltd., 1960.

Strahler, Arthur N., *A Geologist's View of Cape Cod.* New York: The Natural History Press, 1966.

Stratton, Arthur, *The Great Red Island.* New York: Charles Scribner's Sons, 1964.

Strode, Hudson, *The Story of Bermuda.* New York: Harrison Smith, 1932.

Suggs, Robert C., *The Hidden Worlds of Polynesia.* New York: The New American Library of World Literature, Inc., 1965.

Suter, Russell, *Mapping of Geologic Formations and Aquifers of Long Island, New York.* Albany, N.Y.: Bulletin GW-18, 1949.

Synge, John M., *The Complete Works of John M. Synge.* New York: Random House, 1904.

372

Taylor, E. G. R., *The Haven-Finding Art: A History of Navigation From Odysseus to Captain Cook.* New York: Abelard-Schuman Limited, 1957.

Teale, Edwin Way, *Dune Boy: The Early Years of a Naturalist.* New York: Dodd, Mead & Company, 1958.

Teale, Edwin Way, *Journey Into Summer.* New York: Dodd, Mead & Company, 1966.

Teale, Edwin Way, *North With the Spring.* New York: Dodd, Mead & Company, 1966.

Teale, Edwin Way, *Wandering Through Winter.* New York: Dodd, Mead & Company, 1966.

Tebeau, Charlton W., *Florida's Last Frontier: The History of Collier County.* Coral Gables, Fla.: University of Miami Press, 1957.

Teller, Walter (ed.), *Five Sea Captains: Their Own Accounts of Voyages Under Sail.* New York: Atheneum, 1960.

Terres, John K. (ed.), *Discovery.* Philadelphia and New York: J. B. Lippincott Company, 1961.

Thaxter, Celia, *An Island Garden.* Boston and New York: Houghton Mifflin and Company, 1894.

Thaxter, Rosamond, *The Life and Letters of Celia Thaxter.* New Hampshire: Marshall Jones Company, 1963.

The Boston Public Library Quarterly. Boston, Mass.: Published by the Trustees, 1949.

The Century (Illustrated Monthly Magazine). November, 1881, to April, 1882. New York: The Century Co., 1882.

The Journal of Long Island History. The Long Island Historical Society, 1967.

The Poetical Works of Alfred Tennyson, Poet Laureate. New York: The American News Company, N.D.

The Southwest. (By the Editors of *Look.*) Boston: Houghton Mifflin Company, 1947.

The Travels and Essays of Robert Louis Stevenson. New York: Charles Scribner's Sons, 1895.

The Travels of Sir John Mandeville. New York: Dover Publications, Inc., N.D.

Thomas, Dylan, *The Beach of Falesa.* New York: Stein and Day Publishers, 1963.

Thomas, Gordon, and Witts, M. Max, *The Day the World Ended.* New York: Stein and Day, 1965.

Thorarinsson, Sigurdur, *Surtsey the New Island in the North Atlantic.* New York: The Viking Press, 1964.

Thoreau, Henry David, *Cape Cod.* New York: W. W. Norton & Company, Inc., 1951.

Thorp, Margaret F. *Sarah Orne Jewett.* Minneapolis: University of Minnesota Press, 1966.

Throckmorton, Peter, *The Lost Ships.* Boston, Toronto: Little, Brown and Company, 1964.

Tillyard, E. M. W., *The Elizabethan World Picture.* New York: Random House, N.D.

Tinbergen, Niko, *The Herring Gull's World.* New York: Basic Books, Inc., Publishers, 1961.

Todd, Charles B., *In Olde New York.* New York: The Grafton Press, 1907.

Tourism in Greece. Published by "Helleneus." 1967.

Trumbull, Gurdon, *Names and Portraits of Birds.* New York: Harper & Brothers, 1888.

Trumbull, Robert, *Paradise in Trust.* New York: William Sloane Associates, 1959.

Tschorik, Harry, Jr., *Indians of North America.* New York: The American Museum of Natural History, 1952.

Tucker, Terry, *Bermuda and the Supernatural.* Bermuda: Longtail Publishers, 1968.

Twentieth-Century Art From the Nelson Aldrich Rockefeller Collection. U.S.A.: Volk & Huxley, Inc., 1969.

Valentine, D. T., *Manual of the Corporation of the City of New York.* New York: Edmund Jones & Co., Printers, 1863.

Van Bergeijk, Willem A.; Pierce, John R., and David, Edward E., Jr., *Waves and the Ear.* Garden City, N.Y.: Doubleday & Company, Inc., 1960.

Vandercook, John W., *Dark Islands*. New York and London: Harper & Brothers Publishers, 1937.

Van Doren, Mark (ed.), *The Travels of William Bartram*. New York: Barnes & Noble, Inc., 1940.

Verne, Jules, *The Mysterious Island*. Garden City, N.Y.: Doubleday & Company, Inc., 1961.

Verrill, Addison E., *The Bermuda Islands*. New Haven, Conn.: Published by the Author, 1907.

Villiers, Alan, *The Ocean: Man's Conquest of the Sea*. New York: E. P. Dutton & Co., Inc., 1963.

Voe, Thomas F., *The Market Assistant*. New York: Hurd and Houghton, 1867.

Von Tschudi, J. J., *Travels in Peru, During the Years 1838–1842*. New York: George P. Putnam, 1848.

Waddell, D. A. G., *The West Indies and the Guianas*. Englewood Cliffs, N.J.: Prentice-Hall, Inc., 1967.

Wade, Allan (ed.), *The Letters of W. B. Yeats*. New York: The Macmillan Company, 1955.

Wagner, J. L. Robert, and Abbot, R. Tucker (eds.), *Van Nostrand's Standard Catalog of Shells*. Princeton, N.J.: D. Van Nostrand Company, Inc., 1964.

Wagner, Kip, as told to Taylor, L. B., Jr., *Pieces of Eight: Recovering the Riches of a Lost Spanish Treasure Fleet*. New York: E. P. Dutton & Co., Inc., 1966.

Walkinshaw, Robert, *On Puget Sound*. New York: G. P. Putnam's Sons, The Knickerbocker Press, 1929.

Wallace, Alfred R., *Malay Archipelago*. New York: Dover Publications, 1969.

Ware, Richard D., *In the Woods and on the Shore*. Boston: L. C. Page & Company, 1908.

Watson, E. L. Grant (ed.), *Nature Abounding*. London: Faber & Faber Limited, N.D.

Waugh, Alec, *The Sunlit Caribbean*. London: Evans Brothers Limited, 1948.

Welker, Robert Henry, *Birds and Men: American Birds in Science, Art, Literature, and Conservation, 1800–1900*. New York: Atheneum, 1966.

White, Frederic R., *Famous Utopias of the Renaissance*. New York: Hendricks House, 1946.

White, William (ed.), *By-Line: Ernest Hemingway*. New York: Charles Scribner's Sons, 1967.

Whitehill, Walter Muir, *The East India Marine Society and the Peabody Museum of Salem: A Sesquicentennial History*. Salem, Massachusetts: Peabody Museum, 1949.

Wilstach, Paul, *Tidewater Virginia*. New York: Blue Ribbon Books, Inc., 1929.

Wittmer, Margret, *Floreana Adventure*. New York: E. P. Dutton & Co., Inc., 1961.

Wolff, Werner, *Island of Death*. New York: J. J. Augustin Publisher, 1948.

Woodham-Smith, Cecil, *The Great Hunger*. New York: Harper & Row, Publishers, 1962.

Wright, John K., *The Geographical Lore of the Time of the Crusades*. New York: Dover Publications, Inc., N.D.

Wright, Richardson, *Gardener's Tribute*. Philadelphia and New York: J. B. Lippincott Company, 1949.

Wychoff, Jerome, *Rock, Time, and Landforms*. New York: Harper & Row, Publishers, 1966.

Yale, Leroy M., *The Out of Door Library. Angling*. Charles Scribner's Sons, 1896.

Yeats, W. B., *Collected Poems*. London, Melbourne, Toronto: Macmillan, 1967.

Yonge, C. M., *The Sea Shore*. New York: Atheneum, 1963.

Zim, Herbert S., and Hoffmeister, Donald F., *Mammals*. New York: Simon and Schuster, 1955.

Zim, Herbert S., and Ingle, Lester, *Seashores: A Guide to Animals and Plants Along the Beaches*. New York: Golden Press, 1964.

375

Index

Italicized numbers refer to pages containing illustrations.

Seon and Robert Manley

Robert Manley holds an engineering degree from the University of Wisconsin and has completed graduate work in economics and marketing. He is president of Manley Management and Marketing Services, and has been a director of the Society of Professional Management Consultants, a former officer of the Association of Management Consultants, and presently is an active member of the International Steering Committee of Sales and Marketing Executives (SME), the Society for Advancement of Management (SAM), the New York Chapter of the American Nuclear Society, and the Downtown Economists Club in New York. Mr. Manley recently became a founding member and a director of the Institute of Management Consultants.

Seon Manley has been active in the fields of publishing and graphic arts for two decades. She is vice president of her husband's company, where she has worked on all of the firm's communication, marine technology, environmental and ocean engineering projects. Mrs. Manley is the author of a number of books and is well known as a historian for both Fire Island and Long Island. She has written the highly praised *Long Island Discovery* and is editor with her sister, Gogo Lewis, of such varied works as *To You With Love: A Treasury of Great Romantic Literature* and *The Oceans: A Treasury of the Sea World*. Her book, *My Heart's in the Heather* was widely acclaimed on its appearance in 1969.

Seon and Robert Manley's backgrounds combine the world of books and the world of electronics in such a fashion that they speak knowledgeably of the revolutions in marine and shore technologies. They have worked with coastal conservation groups throughout the country. Their obvious love of islands is shown in this present work, as their love of beaches was evidenced in the companion piece to this book, *Beaches: Their Lives, Legends and Lore*, published by Chilton in 1968.

Mr. and Mrs. Manley, and their daughter Shivaun, live in Greenwich, Connecticut, when they are not collecting islands and beaches.

DATE DUE
